Assessing Revolutionary and Insurgent Strategies

THRESHOLD OF VIOLENCE

Paul J. Tompkins Jr., USASOC Project Lead

Guillermo Pinczuk, Lead Author/Editor

Chris Kurowski, Contributing Author

United States Army Special Operations Command
and
The Johns Hopkins University Applied Physics Laboratory

This publication is a work of the United States Government in accordance with Title 17, United States Code, sections 101 and 105.

Published by:

The United States Army Special Operations Command

Fort Bragg, North Carolina

Reproduction in whole or in part is permitted for any purpose of the United States government. Nonmateriel research on special warfare is performed in support of the requirements stated by the United States Army Special Operations Command, Department of the Army. This research is accomplished at the Johns Hopkins University Applied Physics Laboratory by the National Security Analysis Department, a nongovernmental agency operating under the supervision of the USASOC Sensitive Activities Division, Department of the Army.

The analysis and the opinions expressed within this document are solely those of the authors and do not necessarily reflect the positions of the US Army or the Johns Hopkins University Applied Physics Laboratory.

Comments correcting errors of fact and opinion, filling or indicating gaps of information, and suggesting other changes that may be appropriate should be addressed to:

United States Army Special Operations Command

G-3X, Sensitive Activities Division

2929 Desert Storm Drive

Fort Bragg, NC 28310

All ARIS products are available from USASOC at www.soc.mil under the ARIS link.

Published by Conflict Research Group.

First published by USASOC in 2019

CONFLICT RESEARCH GROUP

ASSESSING REVOLUTIONARY AND INSURGENT STRATEGIES

The Assessing Revolutionary and Insurgent Strategies (ARIS) series consists of a set of case studies and research conducted for the US Army Special Operations Command by the National Security Analysis Department of the Johns Hopkins University Applied Physics Laboratory.

The purpose of the ARIS series is to produce a collection of academically rigorous yet operationally relevant research materials to develop and illustrate a common understanding of insurgency and revolution. This research, intended to form a bedrock body of knowledge for members of the Special Forces, will allow users to distill vast amounts of material from a wide array of campaigns and extract relevant lessons, thereby enabling the development of future doctrine, professional education, and training.

From its inception, ARIS has been focused on exploring historical and current revolutions and insurgencies for the purpose of identifying emerging trends in operational designs and patterns. ARIS encompasses research and studies on the general characteristics of revolutionary movements and insurgencies and examines unique adaptations by specific organizations or groups to overcome various environmental and contextual challenges.

The ARIS series follows in the tradition of research conducted by the Special Operations Research Office (SORO) of American University in the 1950s and 1960s, by adding new research to that body of work and in several instances releasing updated editions of original SORO studies.

VOLUMES IN THE ARIS SERIES

SORO STUDIES

Table of Contents

List of Illustrations

Figure 1 courtesy of Dr. Gordon McCormick, Naval Postgraduate School, published in Eric P. Wendt, "Strategic Counterinsurgency Modeling," *Special Warfare* 18, no. 2 (2005): 5.

Figure 2 adapted from Stathis N. Kalyvas, *The Logic of Violence in Civil War* (Cambridge, UK: Cambridge University Press, 2006), 167.

INTRODUCTION

Various authors have noted that violence is often a double-edged sword within combat settings, particularly those involving a resistance movement fighting an asymmetric conflict against the security forces of a stronger incumbent government, with both sides vying for the sympathies of a local population. Atkinson and Kress noted that "on the one hand violence is needed to fight the other side and perhaps deter individuals in the population from supporting the other side, but on the other hand it can turn the population against the source of that violence."[1]

This tension is captured more formally in the equivalent response model originally developed by Dr. Gordon McCormick of the Naval Postgraduate School and elaborated by Wendt.[2]

Wendt noted that, with respect to the use of violence by resistance movements that seek the overthrow of a targeted government, such groups sometimes start small and may follow a certain trajectory in their formative stages. Initially a group may be capable of only small-scale resistance or violent actions, and (if it remains in existence) it may add covert and small-scale guerrilla actions to its repertoire as it grows to potentially culminate with its adoption of conventional tactics. A canonical example is given by the Liberation Tigers of Tamil Eelam (LTTE), which eventually started out in the late 1970s with small-scale attacks in northern Sri Lanka, which was followed by the group's widespread use of irregular tactics in the early 1980s and culminated with its "graduation" to conventional tactics against the Sri Lankan government beginning in the 1990s.

Wendt noted that regardless of the length of a resistance movement's gestation period, it must grow in size to become capable of challenging an incumbent government, and to do so, it must become a judicious user of increasing levels of violence.[a] Figure 1 depicts this

---

[a]   While this report focuses on the use of violence by a resistance movement against a government, one should keep in mind that violence is only one form of contention that groups can employ against a governing authority. There are a multitude of nonviolent forms of resistance to authority, including strikes, protests, sit-ins, and marches, and in a 2008 study of violent and nonviolent resistance campaigns from 1900 to 2006, Stephan and Chenoweth noted that major nonviolent campaigns achieved success 53 percent of the time, compared with 26 percent for violent resistance campaigns.[3]

range of necessary yet tolerable levels of violence.[b] At a minimum, a group must use a minimum level of violence to ensure its relevance and assist with mobilization efforts, and as a group grows in popularity over time, this minimum level may need to increase to ensure that the group's violent activities convey perceptions of strength commensurate with its enhanced status. However, excessive violence may be detrimental and self-defeating if it violates a community's norms surrounding legitimate levels (and targets) of violence or if it brings a crippling counterresponse from the government.

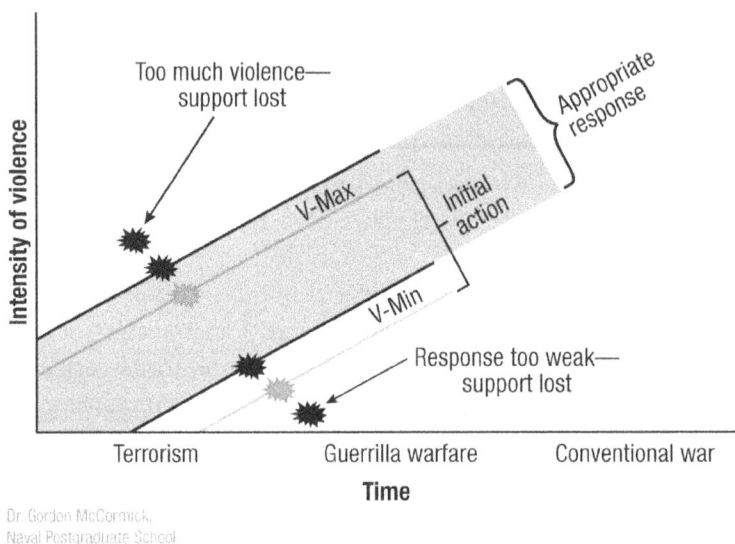

Dr. Gordon McCormick,
Naval Postgraduate School

Figure 1. Equivalent response model.

It therefore follows that, with respect to the use of violence, resistance groups may need to operate within a "band of excellence" in which the intensity of violence is maintained within minimum and maximum thresholds (V-min and V-max in Figure 1). The government also faces upper and lower thresholds of violence, and Wendt noted that for a government to maintain legitimacy and provide security to a host population, the intensity of its response might have to exceed the initial salvo of a resistance group. Hence, both an incumbent government and a resistance movement challenging the government's authority confront

---

[b] As a model, and in particular as a social-scientific model, the equivalent-response model is a simplification of a complex reality, and one may take issue with various assumptions and axioms implicit within the model or with the conclusions of the model (or both). Hence, not all groups that seek to challenge or overthrow a government seek to establish in linear time a conventional capability (examples include the Shining Path in Peru or the Montoneros in Argentina). Furthermore, sometimes groups may come to power even after they have used violence injudiciously. For instance, the Zimbabwe African National Union came to power in 1980 despite having alienated segments of the non-Shona population in Zimbabwe. See Cliffe, Mpofu, and Munslow.[1]

lower and upper thresholds of violence, where too little violence signals weakness and ineffectiveness (see "Response too weak—support lost" in Figure 1), while excessive levels of violence risk alienating the host population (see "Too much violence—support lost"). The fundamental dilemma, of course, is that the lower and upper thresholds of violence and their evolution over time are determined by a local population, and assessments of those levels and their evolution by insurgent or government decision makers are highly subjective and error prone. Speaking of the use of violence by the Irish Republican Army (IRA), Darby noted:

> On the evidence of the IRA's use of legitimate targeting, its denials of unwanted casualties, its exclusion of certain groups from attack and its care to anticipate internal criticism, it is clear that the IRA is aware of the limits of its own community's tolerance. The need to maintain the tolerance has been a major restraint on its escalation of its campaign of violence towards a more genocidal indiscriminate slaughter. The instruments for measuring the community's toleration are not precise ones. The limits are often defined only when they are breached, and the community indicates by the means of communication at its disposal that the violence has gone too far.[5]

A number of recent examples illuminate violations of the threshold of violence, in particular the upper threshold. For instance, in July 2005 Ayman al-Zawahiri, then second in command of al-Qaeda, wrote a letter to Abu Musab al-Zarqawi, a prominent Islamic militant in Iraq, criticizing the latter's reliance on attacks on Iraqi civilians and use of gruesome methods to kill hostages.[6] The al-Qaeda leader realized that such gratuitous displays of violence had alienated Muslim public opinion, thereby threatening the achievement of the group's strategic goals to remove the United States from the Middle East and reestablish the caliphate.

Interestingly, violations of the upper threshold need not be normative. They may come in the form of infringements on the interests of powerful political, economic, and social elites. More specifically, violence and associated activities may endow groups with political influence that cuts against the interests of local elites, who may then seek to mobilize the local population against the initiating group that is using violence to upend local hierarchies.

An example is the decision by Sunni tribal leaders in Iraq's Anbar Province to break with al-Qaeda and side with the United States beginning in 2005. As McCary stated:

> Although media coverage and analysis has focused heavily on al Qaeda's campaign of violent coercion and the supposed improved efficacy of the US military after the arrival of the "surge" brigades, testimony from Iraqis themselves and US military commanders on the ground tells a different story of why the sheikhs chose to change sides . . . it was not the grotesqueness of the violence perpetrated by al Qaeda which caused the change, for Iraq and al Anbar have a long storied history of using violence for political ends. Rather there appear to be two main factors: the Sunni tribal sheikh's own changing perception of al Qaeda's threat to their continued hold on power and the developing US military approach in al Anbar.
>
> Initially, the major threat to the sheikhs was the US military and its imposition of martial law, democratic processes, and support of the new Iraqi government based in Baghdad as the locus of power for the country. All of these elements undermined their traditional position of power in the region and disrupted their ability to control their tribesmen. Later, the sheikhs began to face a similar threat from al-Qaeda itself, which increasingly asserted control in the region through money and violence but also posed a clear and real mortal threat to the leaders themselves.
>
> Eventually, Sunni tribal leaders in Anbar Province deemed al-Qaeda's influence as more of a threat to their continued rule, while US forces were considered to be less and less of a determining factor in the region.[7]

Hence, Sunni tribal sheikhs in Anbar began to see al-Qaeda as its primary long-term threat rather than the United States, with a presence in Iraq that was seen as temporary. Their reaction against al-Qaeda was based not on moral repulsion over the use of violence but rather on the belief that increasing levels of jihadi violence enhanced the power and status of al-Qaeda in the province, which threatened the interests (and lives) of the sheikhs themselves, who sat atop local political, economic, and social hierarchies.

The relevant audiences whose opinions shape the threshold need not be restricted to a domestic population but may also include foreign audiences. For instance, in the 1980s, the population of the southern Indian state of Tamil Nadu, which at the time consisted of approximately fifty million Tamils, sympathized with the plight of their ethnic brethren in Sri Lanka. The need for Tamil electoral support and to keep latent the desires for independence on the part of the Tamil population and elites in Tamil Nadu forced the Indian government to take a leading role in the ethnic conflict in Sri Lanka. However, the LTTE's assassination of Indian prime minister Rajiv Gandhi in May 1991 proved deeply unpopular in Tamil Nadu.[8] In this case, an insurgent movement's use of violence violated the norms of a foreign audience regarding the appropriateness of the target of a violent attack. Although not defeated until 2009, the LTTE apparently overplayed its hand with the assassination, which led to a sharp decrease in support from the Tamil population in Tamil Nadu.[9]

However, sometimes violations of the threshold may be fatal in the short run. In the late 1980s, the *Janatha Vikmuthi Peramuna* (JVP—The People's Liberation Front), a Marxist Sinhalese resistance movement in Sri Lanka, subjected that country to several years of economic sabotage, political assassinations, and insurgent violence in an effort to overthrow the government. Its activities created havoc throughout the Sinhalese areas of the country, and although it put significant pressure on the government, by mid-1989, it had not succeeded in its efforts to destroy it. Growing desperate, the JVP increased the lethality of its attacks on the one institution holding the country together, the armed forces. In early August 1989, it issued a threat over radio to the family members of armed forces personnel, threatening them with death unless their relatives in the armed forces ceased their campaign against the group. In response, a proregime paramilitary group, the *Deshapremi Sinhala Tharuna Peramuna* (Patriotic Sinhala Youth Front), issued the following chilling threat to relatives of suspected members of the JVP:[10]

Dear Father/Mother/Sister,

We know that your son/brother/husband is engaged in brutal murder under the pretense of patriotism. Your son/brother/husband, the so-called patriot, has cruelly taken the lives of mothers like you, of sisters, of innocent little children. In addition, he has started killing the family members of the heroic Sinhalese soldiers who fought with the Tamil Tigers and sacrificed their lives in order to protect the motherland.

> Is it not among us, ourselves, the Sinhala people that your son/brother/husband has launched the conflict in the name of patriotism? Is it then right that you who are the wife/mother/sister of this person who engages in inhuman murder or your children should be free to live? Is it not justified to put you to death? From this moment, you and all your family members must be ready to die!
>
> May you attain Nirvana!
>
> Patriotic Sinhala Youth Front

This move proved fatal, as the armed forces stepped up a preexisting unconventional campaign against the JVP. The campaign resulted in the final decimation of the group, with much of its leadership and membership wiped out by the end of the year.

As can be seen in these examples, the threshold of violence, in particular the upper threshold, is a highly salient issue affecting insurgent groups, incumbent governments, and concerned publics. This report analyzes how violent resistance movements manage the upper threshold. In particular, it focuses on their decision making regarding the use of violence in light of the threshold and on their strategies to give themselves greater leeway to use violence by raising the upper threshold. Because the decision to use violence may be the outcome of a deliberative process that weighs the costs and benefits of different alternative (violent and nonviolent) actions, we begin with a discussion of the applicability of a particular paradigm, rational choice theory, to help us understand the degree to which the decision to use violence is influenced by rationality and deliberate cost–benefit calculations. Following this discussion, we detail the benefits that may accrue to groups that use violence, including compelling concessions from incumbent governments and sending signals to different concerned audiences.

Lastly, we discuss strategies groups employ to raise the upper threshold. Such strategies include provoking an indiscriminate government counterresponse likely to foster anger and a greater tolerance of insurgent violence among a population, the provision of public goods and social welfare by a violent resistance movement to gain popular support, and the use of narratives to help define community standards regarding legitimate acts of violence. Although much of the discussion centers around insights and examples consistent with the equivalent response model, in which a resistance movement confronts a government as both parties try to win the support of a host population, the discussion at times also considers violence used in other contexts, such as insurgent attacks on a population. Additionally, we include examples of how groups have attempted to or have succeeded in raising the upper threshold of violence.

RATIONALITY AND VIOLENCE

As can be inferred from the examples presented in the previous section, insurgent leaders must often make choices regarding the appropriate level and targets of violence. Such decisions regarding target selection and violence intensity can be highly consequential to the group. The difficulty of deliberating through such consequential choices typically imposes significant cognitive demands on the part of insurgent leaders and counterinsurgent (COIN) forces arrayed against them. Wendt noted:

> Insurgent warfare is a thinking man's game in the extreme. To be successful, COIN forces must take the initiative, carefully choose their actions, weigh possible actions against the band of excellence, and anticipate the adversary's reaction. An effective COIN strategy initiates actions that fall within the band of excellence but cause the opponent to react with actions that fall outside of the band.[11]

The language used by Wendt in this passage is highly consistent with what is known within the social sciences as the rational choice perspective. This perspective argues that in important decision-making situations, individuals select choices to maximize gains or "utility" by relying on rational cost–benefit calculations that tie means to ends, irrespective of the nationality, ethnicity, race, gender, religion, or other form of identity of the decision maker. Because irregular warfare scenarios impose a requirement of at least some level of rationality on the part of key decision makers (to include insurgent leaders, government and military officials, and concerned populations), to set the stage for the rest of the report, we now turn to a brief discussion of the relevance and applicability of rational choice theory to decision making in unconventional conflicts.

McCormick used the term *strategic frame* to describe one perspective through which to understand terrorist groups' decision making.[12] This mode of analysis is essentially the application of rational choice theory to the analysis of decisions made by leaders of terrorist groups, and the discussion surrounding the strategic frame itself can be expanded to encompass uses of violence other than terrorist acts.

As noted by McCormick:[13]

> Terrorism, in this [strategic frame] view, is an instrumental activity designed to achieve or help achieve a specified set of long-run and short-run objectives. Like any such strategy, it is forward-looking and "consequential," in the sense that the decisions to use terrorism and the nature of the terrorism that is used are based on the anticipated consequences of current actions. It is also "preference-based," in the sense that alternative courses of action are evaluated in terms of their respective impact on terrorist objectives. The decision to act (or not act), in this view, depends on the answers to four questions: What alternative courses of action are available? What are the expected effects of each of these alternatives? How are these expected effects likely to influence group objectives? What decision rule will be employed to decide among the alternatives? Stripped down to its essentials, the terrorist decision-making process, in the strategic view, is one of constrained optimization. Terrorist organizations attempt to either maximize their expected political returns for any given level of effort or minimize the expected costs necessary to achieve a specified set of political objectives.

Additionally, insurgent leaders contemplating the use of violence must also develop estimates of the likelihood of different potential actions other key actors might take, as well as the impact of those actions. Hence:

> What this implies, for our purposes, is that a terrorist group's decision to act (or not act)—a decision that includes its choice of targets, tactics, and timing—is influenced by the decisions of its opponents, of its political constituency, and of any other actors that influence its strategic environment. As the French geographer Vidal de la Blache observed in 1926, an organization's environment imposes permissive and limiting conditions that shape its incentives, its opportunities, and the short- and long-run consequences of its actions.[14]

Kydd and Walter noted that a government's past behavior regarding its use of violence and concessions forms important information that can be used to formulate beliefs about likely future behavior.[15]

Such "prior probabilities" of likely government action in turn can be updated as a confrontation unfolds.

Pedahzur also emphasized the strategic frame in his analysis of a group's decisions to use terrorism:[16]

> The preference for suicide terrorism on the part of terrorist organizations is a consequence of advance planning and informed assessment of other alternatives, as each one of the organizations calculates how it can further its goals vis-à-vis a number of targeted publics. These three publics are: the stronger opponent, the public the organization seeks to represent, and the local political rivals.

Additionally, speaking of Hamas's decision to begin using suicide terrorism in the early 1990s, Pedahzur noted:[17]

> In their view, violence is a means and not an end and the decision whether to use violence or not and, if so, which tactic should be chosen, is an outcome of the anticipated benefits inherent to each method of operation.

Kuznar noted that rational choice theory has assumed a canonical role over the last century in the field of microeconomics, and he noted that its key elements include the following:[18]

- Collective economic phenomena (prices, national accounts, production efficiency) are the result of the individual decisions of autonomous decision makers.
- Individuals have full knowledge of their preferences.
- Individuals have full knowledge of the resources they have to satisfy their desires.
- Individuals maximize their satisfaction, or utility, by optimally allocating their scarce resources to alternative ends.
- Individuals possess all the capabilities for calculating how to optimally allocate their resources.
- Individuals are concerned only with maximizing their own utility; they are unconcerned with how poorly or well others are doing.

While these key elements pertain to the economic decisions of individuals, they can potentially also be applied to the decision calculus of insurgent leaders (and COIN forces) attempting to make optimal decisions given existing political, economic, social, diplomatic, and military constraints.

Kuznar noted that these key aspects represent idealizations. For instance, individuals (including insurgent leaders) never possess perfect information about everything they need to know to behave optimally, and no decision maker possesses the cognitive complexity to always behave optimally when faced with highly complex problems.[c] Nonetheless, the strategic frame or rational choice paradigm may offer a useful starting point through which to analyze insurgent decisions on violence intensity and target selection. From this perspective, the use of violence is seen as the product of a cost–benefit calculation in which the tactic of violence used against a selected target is deemed as the most effective way of achieving a desired end state. Social scientists sometimes use the phrase *state of the world* to represent a potential future end state, and this term signifies the political, economic, social, diplomatic, and military conditions extant within a political locality at some point in the future. Hence, from a rational choice perspective, the use of violence by insurgent groups may be seen as a rational means for achieving a desired state of the world at some point in the future.

There are various examples that show that a rational choice perspective can be useful to understanding insurgent decision making. We have already seen how the JVP escalated its use of violence in August 1989 against the family members of personnel in Sri Lanka's armed forces. Moore noted that, prior to this event, the group waged a widespread campaign of economic sabotage that featured enforced strikes, disruptions to the transportation and health sectors, and a disruption in economic activity in general, yet it refrained from attacking the country's tea plantations, which were a significant source of foreign exchange.[21] A rational choice perspective may help illuminate the JVP's decision making with respect to targeting.

Specifically, Moore argued that the only plausible explanation for the group's sparing of the plantations was that it feared the public's reaction to the anticipated widespread economic devastation such an attack would generate.[22] The JVP's ideal state of the world entailed an overthrown and replaced government, with society, politics, and economics reorganized along Marxist-Leninist lines. Although the group recognized that one potential tactic to achieve this end state was an

---

[c] In this regard, a more realistic theory that reflects actual decision making is Herbert Simon's notion of "bounded rationality," which argues that decision makers face a number of limitations when making decisions, including cognitive, temporal, and informational limitations. Thus, decision makers typically are unable to make optimal decisions and instead must settle for satisfactory solutions. Hence, decision makers are "satisficers," rather than "optimizers."[19] We note that an excessive reliance on rational choice theory has been criticized by various social scientists, particularly those prone to using qualitative rather than quantitative methods.[20] See the concluding section for a discussion of other decision-making paradigms that may also shed light on the decision to use violence by insurgent groups and governments.

attack that would cripple the country's main exporting industry, before August 1989, the group likely concluded that the consequences of attacking the tea plantations made this tactic too costly to undertake. It was only after the JVP had likely reasoned it had used all other tools at its disposal, and following the issuance of the death threat against the families of armed forces' personnel, which subjected the group to existential duress, that the JVP actually did attack the plantations in the fall of 1989.

The JVP's decision to issue the death threat can also be analyzed utilizing a cost–benefit calculus. Moore noted that the group had scored successes against the government by the summer of 1989 and was looking for a knockout blow.[23] Additionally, it was aware of the unreliability of the armed forces and their lack of interest in fighting the JVP (as opposed to fighting the LTTE), and the group hoped that many members of the army would desert and take their weapons with them to defend their families, thus precipitating a collapse of the institution.[24] Thus, the group likely reasoned that the benefits of issuing the death threat outweighed any potential costs. In this case, its decision regarding target selection (taken to achieve a desired future end state) proved disastrously incorrect. Hence, when considering the applicability of the strategic frame or rational choice theory, it is useful to keep in mind that insurgent leaders (and other decision makers more generally) may be regarded as rational in a weak sense: they judge means by their expected impact on the likelihood of achieving desired ends.[25] Simply engaging in a cost–benefit analysis and attempting to anticipate the reactions of opponents and divine future end states does not guarantee that errors in judgment will not be made.

Another interesting example, also from Sri Lanka, involves the previously mentioned LTTE decision to assassinate Prime Minister Rajiv Gandhi. While in office in 1987, Gandhi essentially imposed a peace treaty that required the Sri Lankan government to make significant concessions to its Tamil minority in return for the disarmament of Tamil militant groups, including the LTTE. The LTTE ultimately did not disarm, and the group waged a two-year insurgency against deployed Indian troops, which were ultimately withdrawn. Although Gandhi was defeated in parliamentary elections in December 1989, the LTTE carried out the assassination in May 1991 over fears he would return to power, reintroduce troops, and enforce the peace treaty, which would require the LTTE to abandon its dreams of an independent Tamil state on the island.[26]

The intense backlash against the assassination, which resulted in the loss of Tamil Nadu as an operational and logistical base,[27] suggests that the LTTE may have miscalculated in its initial assessment of the utility of the operation. Indeed, years later, the group's chief negotiator and theoretician, Anton Balasingham, rued the decision to assassinate Gandhi:[28]

> As far as that event is concerned, I would say it is a great tragedy, a monumental historical tragedy for which we deeply regret and we call upon the government of India and the people of India to be magnanimous to put the past behind and to approach the ethnic question in a different perspective.

One may be forgiven in thinking that a rational choice perspective implies that history is linear or efficient, with decision makers weighing different one-time actions that most effectively shift the state of the world to a more desirable end state, after which history ends. Instead, history may be viewed as a cycle of repeated interactions between different participants, with no clear terminal end state. Addressing Vladimir Putin's strategy toward Ukraine in July 2014, a writer for the *Economist* observed:[29]

> Mr. Putin's apparent interest in an on-again, off-again cycle of ceasefires and negotiations suggests that he would like to lock the conflict in place, legitimizing the so-called people's republics in Donetsk and Luhansk and raising the profile of these rebel commanders ready to take orders from Moscow. Such a war—oscillating between open fighting and political talks, but without ever being completely resolved—would resemble other frozen conflicts around the former Soviet Union. In Abkhazia, South Ossetia and Transniestria, unsettled but largely bloodless conflicts serve to constrain the Georgian and Moldovan governments as well as providing a built-in lever for frequent Russian meddling.

Similarly, as part of an exercise of rational deliberation, decision makers may also place greater value on long-term rather than short-term end states. Leaders of Hamas and other Palestinian insurgent groups may be adopting a time horizon that spans decades into the future, when the demographic balance in Israel, the West Bank, and the Gaza Strip has already shifted in favor of the Palestinians. By this time, international pressure on Israel to reach a settlement in favor of the Palestinians may prove insurmountable. From this perspective, violent and non-violent actions employed by Hamas against Israel could be viewed as efforts by the group to position itself in what it sees as a likely future sometime in the next ten to twenty years.

A number of authors regard non-elites within a population as rational decision makers (in a weak sense) who engage in careful cost–benefit calculations when deciding whether to support a resistance movement or an incumbent government. In a much-cited study, Popkin argued that Vietnamese peasants during the Vietnam War could be viewed as rational problems solvers aware of their own interests and ready to bargain with others to achieve a desired outcome.[30] Mason noted that many analysts regard non-elite behavior as non-ideological, primarily apolitical, and concerned more with short-term damage limitation or benefit maximization than with preferences for particular institutional or ideological arrangements (e.g., the establishment of communism, liberal democracy, or sharia law within a territory) preferred by competing elites.[31] Sometimes non-elites are motivated both by material interests (e.g., economics, security, or well being) and ideological concerns (e.g., identity, nationalism, ethnicity, or religion), but in cases of acute conflict, the former may take precedence. One despondent Palestinian woman whose house was destroyed by Israel during Operation Cast Lead observed, "I will never vote for Hamas. They are not able to protect the people, and if they are going to bring this on us, why should they be in power? If I thought they could liberate Jerusalem, I would be patient. But instead they bring this on us.[32]

Additionally, McCormick observed that, with respect to popular support for violent groups, supportive members of a population can generally be classified as either committed supporters or opportunistic fence sitters who wait on the sidelines to see which side comes out victorious. Generally speaking, the fence sitters are more numerous. He stated:[33]

> Every violent political contest is defined by a "hard core" element on each side, whose members are prepared to "stick to their guns" regardless of which side is ahead. The (effective) support of the majority of the population, by contrast, typically depends on their subjective estimates of each side's prospects.

In addition to forming a subjective estimate of each side's prospects, members of the general population must also estimate the likelihood that they will be subjected to violence if they support one side or another. Speaking of the decision to participate within a protest or demonstration, DeNardo noted that:[34]

> The generalized model of ideological recruiting assumes that potential demonstrators can anticipate the level of repression they will face and estimate the probability that it will be visited upon them. This

probability surely depends on the repressive capacity of the regime—something that is generally known in most places—but also on the number of people who ultimately take to the streets.

As suggested by DeNardo, the likelihood that a demonstrator will be subjected to violence varies inversely with the number of demonstrators. An increase in participation suggests that a movement has greater strength (and therefore a greater capacity to withstand an incumbent government), and through sheer probability, an increase in numbers indicates that any one individual is less likely to be targeted with violence. Although DeNardo's analysis is in the context of deciding whether to participate in a social movement considering an escalation of tactics, it applies even more to whether or not an individual chooses to support a violent insurgent movement against an incumbent government.

In summation, we see that a rational choice perspective suggests that individuals are weakly rational in the sense of being capable of engaging in cost–benefit calculations that tie ends and means and, at times, may offer a useful perspective through which to understand decision making in conflicts. Furthermore, weak rationality is not limited to elite decision making but may also extend to non-elites as well. However, there are other perspectives through which to view insurgent leaders' decision making. McCormick noted that decision making can also be viewed from an organizational frame, in which organizational concerns weigh heavily on the decision to use violence.[35] For instance, he noted that the intense need for secrecy that is a distinguishing feature of terrorist groups has the potential to alienate such groups from their larger political and social settings. This risk also applies to violent resistance movements more generally, and the concern is that, under such conditions, violent resistance movements become more hidebound as their decision making becomes more rigid, closed, and inward looking. Additionally, groups may also be prone to groupthink, in which pressure for decision makers to agree impacts their ability to evaluate options. Lastly, groups may also find themselves in competition with other violent resistance movements for popular support, thus forcing them to engage in violence to win public approval (as opposed to worrying about the loss of support that may come with the use of violence). This may be the case if a public desires revenge against another group or governing authority, and various groups may therefore compete with one another to provide what is essentially a public good. Such a process has been termed *outbidding* and will be discussed later in the next section.[d]

Lastly, a counterpoint to the strategic frame is that group actions at times may not be motivated by the pursuit of a strategic goal but rather by idiosyncratic moral calculations. For instance, Libicki, Chalk, and Sisson suggested that some al-Qaeda actions may be motivated by a moral calculus whereby the group perceives that its actions are sanctified by Allah and conducted in the pursuit of justice, which provides a sufficient reason for using violence.[37,e] We discuss other decision-making paradigms in the conclusion. Despite these objections and qualifications, however, the strategic frame may prove to be a useful starting point as we subsequently consider the decision-making process of insurgent leaders and concerned publics regarding the use of violence by violent resistance movements.

USE OF VIOLENCE

The previous two sections noted the motivations that groups such as the JVP and LTTE had for using violence. In this section, we explore this important topic more generally. On a very basic level, resistance movements use violence to gain concessions by demonstrating a "power to hurt" the other side, and they threaten or use force to achieve a desirable end state. Indeed, if we pick any asymmetric conflict between a resistance movement and a government, we are likely to see the former use violence against the latter, and vice versa, to achieve desirable end states. Examples include the use of violence by Hizbollah and Hamas to compel an Israeli withdrawal from southern Lebanon and the Gaza Strip in 2000 and 2005, respectively; the LTTE's use of violence against the Sri Lankan government to achieve an independent Tamil ethnocracy in northern Sri Lanka; and the Shining Path's violent efforts to remake Peru's economic and political systems along revolutionary communist lines. Often, groups intend to use violence to bring about a change in public opinion to facilitate changes in policy. For instance, Drake noted that the Front de Libération Nationale used bombs to

---

[d]   Another perspective through which to view insurgent decision making is through what McCormick labels the "psychological frame," which emphasizes the impact of intra-psychic processes and individual psychology on decisions to use violence. For instance, some have argued that acts of terrorism are motivated by narcissistic wounds to an individual's self-image and self-esteem and that terrorism serves as an auto-defense mechanism to restore a sense of self-worth.[36]

[e]   McCormick noted that terrorists may wage a "fantasy war" in which they believe they are acting on behalf of a larger community and that their attacks are always defensive and conducted in response to provocations. Such beliefs can function as "autopropaganda."[38]

attack European civilians in Algeria during its 1954–1962 war with France to polarize Europeans from Algerians, thereby making French rule more difficult.[39]

Interestingly, Thomas noted that violence can provide benefits even if its use does not culminate in immediate concessions.[40] Violence can also be used to gain a seat at the bargaining table, which in effect may actually kickstart a peace process.[f] Examining monthly data on civil conflicts in Africa from 1989 to 2010, she found that the likelihood that rebels engage in negotiations (in the next month) when they do not conduct a successful terror attack is 30 percent, yet this probability nearly doubles after ten successful attacks in a month and rises to approximately 88 percent after twenty successful attacks.[42]

Various authors have elaborated on terrorist groups' strategies for using violence, and this discussion can be applied more broadly to violent resistance movements. Kydd and Walter noted five key strategies associated with the use of violence by terrorist groups, including an attrition strategy, an intimidation strategy, a provocation strategy, a spoiling strategy, and an outbidding strategy, with each representing a form of "costly signaling" that can be used to send signals to various audiences.[43] Kydd and Walter defined costly signals as follows:[44]

> Actions so costly that bluffers and liars are unwilling to take them. In international crises, mobilizing forces or drawing a very public line in the sand are examples of strategies that less resolved actors might find too costly to take. War itself, or the willingness to endure it, can serve as a forceful signal of resolve and provide believable information about power and capabilities. Costly signals separate the wheat from the chaff and allow honest communication, although sometimes at a terrible price.

Costly signals communicate levels of resolve, power, and capabilities, and conflict itself can be a form of learning as the prewar period may be characterized by uncertainty and mistaken "prior probabilities" regarding these factors. Indeed, Blainey made the insightful observation that wars begin when states disagree about their relative power and end once they are in agreement.[45] In the case of a conflict between a terrorist movement and a government, Overgaard noted that such conflicts are characterized by "asymmetric information," where initially

---

[f]   Interestingly, the use of violence for this purpose may create a problem of "moral hazard." Thomas[41] noted that after Israel negotiated a cease-fire with Hamas in 2012 after a string of attacks initiated by the group, members of Fatah indicated that Israel's response might make them reconsider their reliance on nonviolent tactics to win concessions.

terrorists and resistance movements are aware of their own capabilities and resolve, but the government is not.[46] Hence, a terrorist or insurgent movement may seek to undertake a large attack to signal its level of resources to soften a potential government retaliation and to also improve its bargaining position.

More generally, resistance movements may believe they need to send costly signals to add credibility to their threats. As Kydd and Walter noted:

> Given the conflict of interests between terrorists and their targets, ordinary communication or "cheap talk" is insufficient to change minds or influence behavior. If al-Qaida had informed the United States on September 10, 2001, that it would kill 3,000 Americans unless the United States withdrew from Saudi Arabia, the threat might have sparked concern, but it would not have had the same impact as the attacks that followed. Because it is hard for weak actors to make credible threats, terrorists are forced to display publicly just how far they are willing to go to obtain their desired results.[47]

Further, they noted:[48]

> To obtain their political goals, terrorists need to provide credible information to the audiences whose behavior they hope to influence. Terrorists play to two key audiences: governments whose policies they wish to influence and individuals on the terrorists' own side whose support or obedience they seek to gain.

The same as well can be said for violent resistance movements more generally. As previously noted, the different strategies identified by Kydd and Walter represent different forms of costly signaling, and we highlight their key aspects:[49]

- Attrition Strategy
  - Objective: Signal strength and resolve; impose costs on incumbent to yield to rebel demands
  - Target: Incumbent force (i.e., existing gov't or foreign occupying power)
- Intimidation Strategy
  - Objective: Discourage collaboration with and decrease legitimacy of incumbent authority
  - Target: Host population and incumbent targets and institutions

17

- Provocation Strategy
  - Objective: Compel indiscriminate response by incumbent; radicalize population against incumbent
  - Target: Incumbent
- Outbidding Strategy
  - Objective: Win popular support through escalating violence
  - Target: Incumbent
- Spoiler Strategy
  - Objective: Scuttle potential peace deal between moderates and incumbent
  - Target: Incumbent

Attrition Strategy

Within an attrition strategy, violence is used to signal strength and resolve and to convey that it is in the best interests of the incumbent government to yield to the demands of the resistance movement. The objective is to impose high costs on the target for its policies, thereby lending credibility to threats to inflict future costs.[50] This was essentially the strategy of the North Vietnamese and Viet Cong against the United States during the Vietnam War, and Kydd and Walter noted that other examples include the revolts against the British by the Greeks in Cyprus, the Jewish population in Palestine, and the Arabs in Aden during the end of the British Empire.[51]

As part of an attrition strategy, groups may also use violence and target selection to communicate resolve to potential supporters. Nemeth noted that attacks on civilians send a signal of resolve to a government by signaling a willingness to violate norms of warfare. At the same time, such attacks signal to potential recruits that the organization is committed to absorbing the costs and repercussions associated with attacks on civilians.[52]

Additionally, groups may use violence to signal resolve and capabilities without intending to secure concessions (or gain a seat at the bargaining table) in the short term. In the January 2015 exchange of attacks between Israel and Hizbollah, it appeared that both sides intended to communicate resolve and strength without actually precipitating a war. One analyst stated that "it's a very delicate game, because both sides want to respond hard enough that they're not perceived as weak, but not too hard to start a war,"[53] while a diplomat observed that "to me, the whole thing was calibrated to say, 'You did your thing, we did our thing.' "[54]

Although violence is an inherent part of an attrition strategy (as well as other strategies), sometimes violent resistance movements and incumbent governments develop informal "rules of the game" to regulate its use, such as restrictions either on attacks on civilians or attacks in certain territories. In such cases, one side may feel that it needs to use violence if the other side crosses a red line by violating an informal restriction.[g] For instance, in the previously mentioned fighting between Hizbollah and Israel, Hizbollah believed that Israel violated informal rules by attacking Hizbollah inside Syria. As one official with extensive contacts with the group stated, "Israel crossed a red line, and if Hezbollah did not react, Israel will not stop . . . [the attack] shows that Hezbollah's confrontation is with Israel, so it can get back its respected position in the Arab world."[56] This quote parallels the earlier point regarding signaling third parties, such as potential supporters.[h]

Intimidation Strategy

Violence can also be used as part of an intimidation strategy to coerce a population and undermine the legitimacy of the government by punishing those individuals who support or collaborate with the government and abide by its rules. As Kydd and Walter noted, an intimidation strategy "works by demonstrating that the terrorists have the power to punish whoever disobeys them, and that the government is powerless to stop them."[58] Byman noted that attacks against a population can compel cooperation with a resistance movement even if the public does not support the goals of the group. At a minimum, he noted, a group needs to convince a population not to denounce it, which can be accomplished by making the population fear violence.[59]

Attacks launched against pro-government citizens or agents of the government, including mayors, police, prosecutors, or political officials, or against government services and economic institutions are designed to discourage collaboration with the government and signal that attacks against a population cannot be prevented. Such attacks serve to drain legitimacy from an incumbent government and encourage individuals to seek alternative security arrangements with the

---

[g]   For a general discussion of informal rules of the game and other "wartime political orders," see Staniland.[55]

[h]   Hizbollah may also have been trying to send a signal to the Islamic State of Iraq and the Levant, which appointed a Lebanese emir in late January 2015.[57]

resistance movement.[60, i] Such was the strategy of the JVP in the late 1980s as it unleashed a wave of attacks against Sri Lankan political and economic institutions and against political figures in both the ruling and opposition parties.[62]

Violence used as part of an intimidation strategy may also be intended to achieve demographic or political goals. For instance, attacks against a population can be intended to intimidate a population to leave a region. Such migration can make a region more politically viable and can forestall the establishment of ethnic enclaves, which can potentially house fifth columnists and subversive groups. Additionally, the existence of ethnic enclaves can be used as an excuse by co-ethnics in surrounding territory to launch a war for their defense and liberation.

For instance, ethnic transfer was clearly the goal of LTTE attacks on Muslims in northern Sri Lanka, to encourage Muslim emigration to other regions of the island and thus ensure the political viability of a separate Tamil region in northern Sri Lanka.[63] Byman noted that population transfer was also the goal of the 1947 attack by the Irgun and Stern Gang that killed 254 Arab inhabitants of the village of Deir Yassin.[64] The attack was designed to frighten Palestinian Arabs from villages, thereby clearing territory for Jewish settlement. Menachem Begin, the founder of the Likud Party in Israel and a former prime minister of the Jewish state, observed that "the massacre was not only justified, but there would not have been a State of Israel without the victory at Deir Yassin."[65] Ethnic cleansing was also widespread in Baghdad after the US invasion and in the former Yugoslavia in the 1990s. Such a strategy of intimidation is designed to "win the census" and establish facts on the ground before eventual peace talks, which may decide to set geographic boundaries based on the existing spatial distribution of different population groups once fighting has ended.[66]

Additionally, hard-liners within a resistance movement may wage attacks against moderates who negotiate with an incumbent government. Such negotiations risk reaching a resolution short of the demands of hard-liners, thereby calling into question previous sacrifices made with the intention of achieving maximalist goals, and even calling into question the existence of a hard-line faction itself because the latter may lose its *raison d'être* in the aftermath of a compromise. This appears to have been the motivation of LTTE attacks against moderate Tamil politicians who negotiated with the Sri Lankan government in an effort to solve the ethnic conflict in Sri Lanka. For the LTTE, moderate politicians represented careerists and sellouts who threatened to cheapen the sacrifices made by insurgents in the effort to achieve an independent

---

i For additional insights into why resistance movements may use violence against a population, see Weinstein.[61]

state, and these attacks were intended to discourage efforts at reaching a settlement with the government.[67]

Provocation Strategy

Insurgent groups may also adopt a provocation strategy, whereby the use of violence against an incumbent government is intended to draw an indiscriminate attack against a host population, thereby fostering radicalization. Under such a strategy, a violent resistance movement seeks to turn an incumbent government's force advantage against it. Speaking of terrorism, McCormick noted:[68]

> Terrorists, as a general rule, begin the game with the ability to see their opponents but a limited ability to attack what they see. The state, by contrast, begins the game with a much greater ability to attack what it sees but a limited ability to see what it wishes to attack. Terrorist groups enjoy an information advantage; the state enjoys a force advantage. This simple asymmetry is a defining feature of any contest between a state and an underground competitor. A strategy of provocation is designed to take advantage of the underground's information advantage and turn the state's force advantage against it by provoking the regime to strike out indiscriminately at targets it cannot see.

Additionally, as Kydd and Walter noted, in some cases, the leadership of a terrorist organization (or violent resistance movement more generally) is more hostile to an incumbent regime than is the general population. Thus, a provocation strategy, by goading a government into an indiscriminate response, is designed to convince a population that the target government is so irredeemably evil that it needs to be replaced or at least compelled into conceding concessions favorable to the group that used violence to encourage the disproportionate reaction.[69]

How does an indiscriminate response accomplish this? Bueno de Mesquita and Dickson noted that an incumbent government's counterresponse to violence provides information to a host population regarding the government's "type," which in this context pertains to whether or not a government is interested in the welfare of an aggrieved population that is host to a violent resistance movement.[70] By generating high casualties and significant economic costs, not only among members of a violent resistance movement and its supporters but also among the general population, the aggrieved population is likely to formulate an updated subjective probability on the nature of the target government, concluding that it is not interested in the population's welfare. Under such conditions, the population may assign a higher updated probability to future end states entailing widespread economic damage caused by a targeted government. Combined with moral repugnance at the government's apparent lack of concern about the safety and welfare of noncombatants, the population may then decide to flock toward the violent resistance movement.

Before using violence to wage a provocation strategy, the leaders of a violent resistance movement are likely to formulate a subjective probability regarding the likelihood that a government will respond with a discriminate or indiscriminate campaign against the group. Within the context of the equivalent response model, a violent resistance movement may wage a provocation strategy if it believes that its use of violence will remain within the band of excellence and if it assigns a high probability that the government will respond with a level of indiscriminate violence that exceeds the upper threshold of violence.

Kalyvas noted that a population's normative reaction to being targeted indiscriminately stems from individuals' anger at being attacked independently of what they did or could have done, which they perceive as deeply unfair. This perception makes a targeted population willing to support risky and violent actions in response.[71] One Guatemalan peasant described this conversion of fear into anger amid indiscriminate violence:[72]

> This was so heavy, so heavy. You were disturbed, you wanted to have some way of defending yourself. The feeling emerged—it wasn't fear but anger. Why do they come persecuting if one is free of faults, if one works honorably? You felt bad, well we all did. Grief but also anger.

One way in which indiscriminate violence feeds anger, as Kalyvas noted, is through the establishment of a distorted incentive structure in which compliance with the demands of an incumbent government against assisting a violent resistance movement may be as unsafe as noncompliance.[73] Members of a violent resistance movement may be no more threatened by punishment than noncombatants, and under such conditions, the general populace may believe it probable that the government's desired end state is the annihilation of the targeted population.[74]

In addition, the populace may flock to a violent resistance movement if it offers the possibility of protection from such an indiscriminate government campaign. A resistance movement may anticipate this type of shift, as one Viet Cong document pointed out:[75]

> The party was correct in its judgment that government doctrine . . . would drive additional segments of the population into opposition, where they would have no alternative but to follow the Party's leadership to obtain protection.

Once a government engages in indiscriminate violence, a violent resistance movement can then engage in "armed propaganda," developing a narrative that depicts the targeted government as insensitive to the needs of the population and therefore as an illegitimate governing entity. In some conflicts, history provides a ready-made ontology with components and relationships that populate a discourse. Such a history will provide a set of grievances replete with a collection of actors and interactions among them, and these may form the building blocks that constitute a narrative a group uses to mobilize a population. For example, the Shining Path's narrative emphasized the racism and economic and political marginalization that befell Peru's indigenous population after the arrival of the Spanish centuries before. By provoking indiscriminate government counterresponses to its violence, the group was able to update and refresh this narrative by emphasizing new themes related to the government's illegitimate uses of force, thereby shifting the emphasis from historical injustice and displacement to current actions taken by the government.[76]

Outbidding Strategy

Another strategy identified by Kydd and Walter is an outbidding strategy. A violent resistance movement may adopt this strategy if it appears that a host population may actually prefer the use of violence against a target government. An example is the Palestinian society's preference for suicide bombings after the buildup of frustration with the Oslo peace process in 1999. Mia Bloom observed that Palestinian support for suicide operations against Israelis was less than 30 percent between May 1997 and March 1999, and by the summer of 1999, support for Hamas had fallen to 10 percent.[77] However, a combination of factors, including a decline in Palestinian faith that the peace process would deliver an independent state, the corruption exhibited by Yasser

Arafat's Palestinian Authority,[j] and the failure of the Palestinian Authority to improve the daily lives of most Palestinians, led to an increase in support for suicide operations. From December 2000 to October 2003, Palestinian support for suicide operations exceeded 60 percent, with support reaching 85 percent in September 2001.[79]

Within the context of the discussion in the prior section on rationality and violence, Palestinian support for suicide bombings suggests that Palestinians assigned a low subjective probability of achieving their desired end state of an independent Palestine through negotiations. Rather, as Mia Bloom noted, they began to view violence as their only option for achieving independence, and under such conditions, various Palestinian groups began to compete and "outbid" each other in the provision of violence to boost mobilization and increase their share of support among the Palestinian population.[80] After a series of suicide bombings in 2000, support for Hamas increased to more than 70 percent,[81] and during this period, various other groups, including the Democratic Front for the Liberation of Palestine, the Popular Front for the Liberation of Palestine, the al-Asqa Martyrs' Brigade, and a new group, an-Nathir (The Warning), began to use the language of holy war and to rely on suicide bombings, even to the point of claiming responsibility for attacks likely carried out by other groups.[82,k]

In this case, the Palestinian strategy was motivated by both outbidding and a desire to provoke Israel, and there is some empirical evidence suggesting that the radicalization of Palestinian opinion following Israeli attacks displayed noticeable temporal patterns and, in particular, decayed over time. Examining longitudinal data on Palestinian public opinion during the second Intifada, Jaeger et al. found

---

[j] The corruption of the Palestinian Authority under Yasser Arafat did not escape the notice of Israel's political class. Speaking as a private citizen during a pro-Israel rally in Washington, DC in April 2002, Benjamin Netanyahu referred to the West Bank as "Arafatistan."[78]

[k] In addition to interorganizational dynamics (i.e., outbidding) fostering the use of violence, there may also exist intraorganizational factors that promote the use of violence against a government or unarmed civilians. For instance, we have already seen how Nemeth argued that groups may use violence against civilians to communicate resolve to governments or potential recruits. Such messages, of course, can also be used to solidify internal cohesion within a group. Additionally, Abrahms and Potter suggested that when groups have decentralized leadership structures that delegate tactical decisions to lower-level members, the latter may have various incentives to attack civilians. For instance, lower-level members may seek to outbid other members and raise their profiles within an organization by attacking civilians. Additionally, lower-level members may lack the organizational clout to marshal the resources needed to attack hardened targets and thus may choose to attack softer targets instead. Elsewhere, Abrahms has also pointed out (see footnote 17) that groups that heavily rely on terrorism against civilians rarely achieve their political objectives, thereby leading to the obvious question of why they engage in it. As this discussion suggests, intraorganizational factors may offer clues regarding this important paradox.[83]

that Palestinian fatalities caused by Israel lowered Palestinian support for negotiations and also for the more moderate Fatah faction (as opposed to Hamas) within one month of the fatalities' occurrence, but this effect completely dissipated after three months.[84] In a separate study, the authors found that increases in support for Hamas or Fatah as these groups pursued an outbidding strategy did not come at the expense of the other party. Specifically, Hamas's gain in support from its use of violence during the second Intifada came at the expense of other Islamist groups, such as Palestinian Islamic Jihad, while the increase in support for Fatah after its use of violence came at the expense of support for secular groups, such as the Democratic Front for the Liberation of Palestine and the Popular Front for the Liberation of Palestine. Additionally, the authors found that support for either Hamas or Fatah was unchanged when the other faction engaged in violence.[85] These results seem to suggest that, at least in this example, when groups compete with each other while implementing an outbidding strategy, a populace may evaluate a particular group relative to those of a similar "type."

Spoiler Strategy

The final strategy identified by Kydd and Walter is the spoiler strategy. With this strategy, extremists within an opposition movement use violence to scuttle any attempts at improved relations between moderates on their side and the targeted government. As the authors observed, the use of violence by extremists exploits lingering uncertainty and mistrust between moderates and the government and is designed to convince the targeted government that moderates are not willing or able to stop violence against the government.[86] Asymmetric information may also be a factor because a targeted government may not be able to observe the extent of actions taken by moderates to curtail violence by extremists and may therefore judge the sincerity of the moderates according to whether violence occurred.

Kydd and Walter cited various examples of a spoiler strategy, including violence by Hamas during ratification and implementation of the Oslo peace process in the 1990s; violence before Israeli elections in 1996 and 2001, during which the more dovish Labor Party was in power (with violence intended to produce an Israeli preference for the more hawkish Likud Party, which was skeptical of the Oslo peace process); and the use of violence after Arafat's electoral victory in 1996. This electoral success cemented the idea that Arafat was a powerful leader within the Palestinian territories, and so the use of violence against Israelis was intended to signal that Arafat was able but unwilling to crack down on violence against Israelis and therefore could not be trusted as a peace partner.[87]

Although Kydd and Walter did not privilege it with its own category, the mobilization of a population factors highly in calculations of the use of violence by resistance movements.[1] In particular, violence can be used to overcome what is known as the "collective action problem." In the 1960s, Mancur Olson noted that it does not necessarily follow that rational, self-interested individuals will act in support of a collective good or common interest, even if doing so will better their situations.[89] This will be the case if acting collectively in pursuit of a common interest is not individually rational, and this phenomenon came to be known as the "collective action problem."

Importantly, a collective action problem may lead to the suboptimal provision of public goods (especially as a group becomes larger) because individual members, believing that the collective good will be provided anyway by the efforts of other members of the group, have an incentive to shirk their responsibility to contribute toward the provision of the public good. Such a strategy of "free riding" may be individually rational because often the consumption of a public good is nonexcludable and available to all, not just to those who took action and paid the costs for its provision. If many members of a group were to behave in this manner, such collective behavior may translate into the suboptimal provision and even nonprovision of the collective good.

How might the collective action problem help explain the mobilization challenges insurgent movements face? One can regard the desired outcome of an insurgency—such as a revolutionary socialist government, a territory under sharia law, a liberal democracy allied with the West, or an ethnocratic enclave carved out of a multinational empire—as a public good requiring self-sacrifice among a relevant population. However, as McCormick and Giordano noted, insurgency is an inherently risky business that, in most cases, offers little possibility of success.[90] More specifically, in the initial stages of a rebellion, insurgent groups tend to be small and less powerful than the state they are fighting. Under such conditions, the collective action problem is often insurmountable, as most potential supporters (correctly) judge that the likelihood of success is remote, leaving only a small number of die-hard activists willing to risk their lives for a desired collective good. For this reason, most opposition groups die young.

How do groups that go on to challenge the state for supremacy overcome this initial mobilization dilemma? One such strategy, as noted

---

[1] The section regarding the use of violence to foster mobilization is adapted from Agan's study on narratives.[88]

by McCormick and Giordano, is to wage violence against highly symbolic icons of state power. If successful, such attacks generate violent images that, whether factually correct or not, attest to the growing relative strength of the insurgents.[91] The process by which this happens, and the role of violence in effectuating this process, is noted by the authors:[92]

> Violence . . . is used as an instrument of armed propaganda. The objective is to advertise the existence of an emerging opposition, raise popular consciousness and define the terms of the struggle. As Thomas Thornton has suggested, incumbents typically enter an insurgency in a natural state of political "inertia," even in the absence of significant popular support. The insurgents, for their part, begin the game as outsiders, an alien political force which "the organism of society will be predisposed to cast out." Before the opposition can even begin the process of building a base of popular support it must first be able to disrupt the system's inertial stability. "In order to do this, the insurgents must break the tie that binds the mass to the incumbents" by removing "the structural supports that give [the system] its strength." These actions, as Thornton goes on to explain, will gradually sever the socio-psychological bonds that tie conditional elements of the population to the state and force them to choose between a disintegrating status quo and an emerging opposition. This cannot be achieved with words; it can only be achieved with violence.

If highly symbolic violent acts are repeatedly carried out, the populace may then be forced to choose between supporting what it perceives as a crumbling status quo and supporting an opposition that appears to be growing in strength. Under such conditions, the majority of the population may support the latter to partake in any spoils of victory and to avoid being on the receiving end of any score-settling violence that may result from having supported the losing side (or not fully supported the winning side).

Additionally, in the case of ethnic civil wars, the use of violence can promote mobilization along ethnic lines and even foster identity formation and maintenance. Indeed, Byman noted that violent resistance movements that are ethnically based and seek independence often face the task of forging a nation (i.e., an "imagined community" of people who perceive some common connection on the basis of some shared characteristic)[m] in addition to building an ethnocracy, or an ethnically based state. He stated:[94]

> The ethnic terrorist group begins the struggle by strengthening ethnic identity. In this effort the terrorist often faces an uphill struggle: region, tribe, sect, family, and state are all rival sources of identity for individuals. So the first task for the terrorist is to make ethnicity politically salient for the larger ethnic community.

Even if unsuccessful, violence can draw attention to identity issues and highlight distinctions between ethnic groups,[95] a process that is magnified when the state response also involves violence:[96]

> Even more beneficial to identity creation than terrorist violence, however, is the state response to violence. State repression creates identities and mobilizes individuals. The Sri Lankan government response to LTTE violence has given rise to the perception that the Sri Lankan state and army act only in the interest of the Sinhalese. Often, weak identities become politically salient when outsiders create an awareness of them. The Basque separatist group ETA made this goal explicit: it sought to force the government to lash out blindly and create a backlash that would increase popular support for the guerrillas. Similarly, the Irgun sought to conduct operations against the British that would compel British security forces to intern, interrogate, and otherwise harass the Jewish community as a whole.

Such communally focused state repression can help activate and politicize latent identities, as noted by one Bosnian schoolteacher during the Yugoslav conflict in the mid-1990s:[97]

> We never, until the war, thought of ourselves as Muslims. We were Yugoslavs. But when we began to be murdered because we are Muslims, things changed. The determination of who we are today has been determined by our killers.

---

[m] Anderson defined a nation as an "imagined community" because "the members of even the smallest nation will never know most of their fellow-members, meet them, or even hear of them, yet in the minds of each lives the image of their communion."[93] The same can be said of ethnic groups.

This section highlighted the multifaceted motivation for the use of violence and its impact.[n] We now turn to analyzing how resistance movements navigate the upper threshold of violence, in particular how such groups can raise this threshold.

MANAGING THE THRESHOLD OF VIOLENCE

The strategies groups use to manage the threshold of violence can potentially provide them with greater operational and tactical leeway. The preceding two sections identified a number of strategies that groups can employ, including provoking an indiscriminate counter-response by a government and using narratives to encourage a host population to adopt a higher tolerance for violence. Before elaborating on these general strategies that may apply across insurgent groups, it may be useful to explore some of the specific tactics groups have used to manage the upper threshold of violence.

For decades, the IRA waged a violent campaign to detach Northern Ireland from the United Kingdom to form a united republic incorporating all of Ireland. The IRA employed a variety of techniques against British and loyalist forces, including bombings and assassinations (as well as "kneecappings"). Darby noted that the group monitored the Catholic community's reaction to its use of violence and that it was keenly aware of the community's tolerance for violence:[99]

> A review of IRA statements clearly demonstrates its awareness of the need to monitor the response of the Catholic community to its actions. Most attacks, such as bombings of economic targets, were announced without comment or justification. Whenever additional care is taken to justify or explain a particular incident, as when the families of serving soldiers were targeted, it was almost always to a background of real or anticipated internal opposition.
>
> On the evidence of the IRA's use of legitimate targeting, its denials of unwanted casualties, its exclusion of certain groups from attack and its care to anticipate internal criticism, it is clear that the IRA is aware of the limits of its own community's tolerance.

---

[n] Another motivation for the use of violence is economic. Groups may use violence to assist with fund-raising, in particular fund-raising from external sponsors, and to raise funds through looting. This report will not address this type of motivation for violence, but to learn more about economic motivations for violence, see Weinstein and Collier and Hoeffler.[98]

How did the group attempt to manage and influence the upper threshold of violence? The previous quote gives several clues. One of the main ways was through "legitimate targeting," or declaring certain groups, institutions, or interests in Northern Ireland legitimate targets for attack.[100] Such declarations were often published in *An Phoblact*, the weekly publication of Sinn Fein, the political arm of the IRA. Darby noted that, between 1970 and 1989, there were 218 references to legitimate targets in *An Phoblact*, with most directed at economic targets (37.6 percent), state institutions (14.2 percent), "criminals" (13.8 percent), and informers (7.8 percent). Interestingly, members of the army and police accounted for just 7.3 percent of cases.[101] By defining what constitutes a legitimate target and publishing this determination in its flagship publication, the group was clearly attempting to raise a population's upper threshold of violence (particularly with respect to target selection rather than level of violence) to a level that would facilitate operations and tactics.

Another tactic was to give sufficient warning of a possible attack on a target, particularly if the target was fungible in the sense of representing civilian infrastructure that can have a military use in certain contexts. For instance, Darby noted that in 1972 the IRA took particular care to justify potential attacks on an army post in the Royal Victoria Hospital (RVH), as suggested by the following warning issued in *An Phoblact*:[102]

> On several occasions since the resumption of offensive operations by the IRA, several streets in the Lower Falls area have come under heavy attack from British Army snipers operating from the roof of the School of Dentistry in the RVH. Until now we have been loath to take retaliatory action because of the proximity of the hospital. If this sniping continues however, action will have to be taken in the interests of the local people.

Darby noted that this threat was widely criticized by the media, which was dismissed by *An Phoblact* as "much wild publicity," with the newspaper publishing a photo purporting to show the army operating from the hospital.[103] Interestingly, this example illustrates that "asymmetric" and "private" information may complicate efforts to manage the threshold. That is, if a group obtains information through intelligence collection, for example, indicating that a civilian target is being used for military purposes, and if such information is not widely known throughout a population, the group may need to take extra actions to justify its target selection (while at the same time protecting its sources and methods).

Other tactics used by the IRA to influence a population's tolerance for violence included taking actions in addition to "legitimate targeting" to lend an air of legitimacy to its use of violence. For instance, violent resistance movements often want to be seen as maintaining a monopoly on the use of (legitimate) violence, and so they may question the legitimacy of rival groups' use of violence, as occurred when the IRA dismissed rival republican activists as "gangsters," "criminals," and "enemies of the people."[104] The implied message sent to the population was that the IRA's use of violence was legitimate, whereas that of other groups was illegitimate.

Another tactic is to give the impression that some type of legal and due process was followed before the threatened or actual use of violence. For instance, the IRA issued the following statement in 1975 regarding the threat to use violence against an informer:[105]

> In the light of this and other unpublished information obtained from him, it was decided that he could not be allowed to remain in the community. Only two ways were open to ensure his removal.
>
> 1. That he be shot and thus permanently separated from the community, or,
>
> 2. That he be exiled from Ireland, so that he could no longer be responsible for the deaths, shootings and imprisonment of other Irish people.
>
> After due consideration of his age, disposition and the state of truce, we decided on the latter penalty— EXILE. He was given 48 hours to leave the country, after which he will be immediately shot on sight.

Lastly, another tactic the IRA used to legitimate its use of violence was to direct attacks at criminals and other antisocial elements, thereby signaling that its use of violence played an important role in the provision of a critical nonexcludable public good (i.e., security). For instance, the IRA used or threatened to use violence against burglars, car thieves, sex criminals, drug dealers, peddlers in stolen goods, and even young "joyriders" who stole cars and caused disruption in many areas.[106] As another example, the JVP waged a similar campaign in Sri Lanka in the 1980s, taking actions against drug dealers and illegal liquor vendors, and the group also attempted to crack down on prostitution.[107]

Now that we have seen how the IRA managed the threshold of violence, what are some of the more generalized strategies that may apply across different resistance movements? As previously mentioned, one potential option available to violent resistance movements is a strategy of provocation, whereby attacks against a government are designed to provoke an indiscriminate counterresponse that causes significant damage to noncombatants. Bueno de Mesquita identified various competing effects of a government crackdown on terrorist mobilization, and these effects apply more broadly to the adoption of a provocation strategy by a violent resistance movement. Specifically, a government counterresponse decreases the ability of a violent resistance movement to carry out attacks, which makes mobilization less attractive to a population; conversely, it generates ideological opposition to a government and imposes economic costs on a population, both of which foster mobilization and help overcome the collective action problem.[108] Before initiating action, a violent resistance movement must estimate a subjective (prior) probability on the likelihood of each of these effects, so we address each at greater length.

At a minimum, before using violence, a resistance movement must estimate a target government's likely response and the ensuing damage it will inflict on the organization. If a movement's leadership attaches a high subjective probability to a state of the world that does not materialize (i.e., a moderate or otherwise manageable government response or outright capitulation to the group), then its existence may be at stake, as was the case when the JVP misjudged the response of the Sri Lankan armed forces when it issued the death threat to family members of serving personnel. Kalyvas noted that Argentine leftists committed a similar miscalculation by planning a campaign of violence to provoke the government to respond indiscriminately in the hope that the response would generate a level of dissatisfaction sufficient to launch a revolutionary process. As Kalyvas noted, Argentine leftists were correct in their judgment regarding the likelihood of an indiscriminate response but incorrect in judging their ability to manage it, as they were eliminated in the process.[109] Hence, incorrect judgments regarding existential factors can be fatal, and a host population may have no choice but to side with the indiscriminate actor if the organization that waged a provocation strategy is eliminated in the process.[110]

The second potential outcome of a crackdown is heightened ideological opposition to the government, which increases radicalization and boosts mobilization in support of the resistance movement. This is the main intent of a provocation strategy, and as noted in the previous section, this response is associated with the intense feelings of anger and unjust victimization that can follow a disproportionate government response. Condra et al. decomposed this potential response into a number of different potential effects and suggested that civilian casualties caused by an incumbent government consist of short-run "information" and "capacity" effects and also longer-run "propaganda" and "revenge" effects.[111]

The authors define a revenge effect as the desire to enact retribution after harm has been committed against an individual or against his or her family, friends, or neighbors. A propaganda effect exploits feelings of humiliation and anger and mobilizes people against an incumbent government or foreign force in the absence of direct harm against themselves, family, or close associates. Civilian casualties may also discourage a host population from sharing information regarding the identity and location of insurgents with an incumbent government or foreign force, out of either anger or fear for their own physical security; the authors label this as an information effect.

Lastly, an increase in civilian casualties, if they are also accompanied by successful attacks on a resistance movement, may degrade the ability of the resistance to carry out successful attacks against a target government (as previously noted); the authors define this as a capacity effect. Analyzing data collected by the Civilian Casualty Tracking Cell maintained by the International Security Assistance Force in Afghanistan, the authors found evidence to support the revenge effect, as they estimated that civilian casualties from a typical incident generated by counterinsurgents were responsible for an additional incidence of violence in an average-sized district over the following six weeks.° Interestingly, the authors found no evidence that out-of-area casualties led to an increase in violence against the International Security Assistance Force, nor did they find evidence in support of the information and capacity effects. Although these results cannot be generalized outside of their specific context, they seem to indicate that increases in a population's threshold of violence may be primarily limited to those individuals directly affected by a government's counterresponse. This suggests that leaders of resistance movements may be able to elevate the upper threshold of violence if they are correct in a belief that a government counterresponse will directly affect a large portion of a populace.

The third potential outcome of a provocation strategy is increased mobilization against a government brought about by the reduction of opportunity costs caused by an indiscriminate response. What this means is that a population may have incentives to join a resistance movement if an indiscriminate campaign causes widespread damage that eliminates income-earning opportunities within the economy. By making it less likely that potential recruits will be forgoing profitable employment opportunities by joining a resistance movement, the

---

° More specifically, the authors found that an International Security Assistance Force-generated incident within an average-sized district of 83,000 led to one additional significant activity over the next six weeks.[112]

former may be more likely to take up arms, especially if doing so offers opportunities to enhance their security or material prospects.

Various authors have provided empirical details of the economic cost of resistance activities on societies where resistance movements are located and on the potential impact of the concomitant reduction in opportunity costs. Examining data from the Palestinian Labor Force Survey, as well as data on the number of suicide attacks during the second Intifada, Benmelech, Berrebi, and Klor found that a success-ful attack was followed by an immediate 5.3 percent increase in the unemployment rate for the district of origin of the suicide bomber (relative to the average Palestinian district-level unemployment rate)[p] and led to a 20 percent increased likelihood that average wages in the district of origin would fall in the following quarter. Additionally, the authors found that a successful attack reduced the number of Palestin-ians working in Israel by 6.7 percent (relative to the average percentage of Palestinians working in Israel for all districts).[114] Additionally, Abadie and Gardeazabal estimated that a 10-percent gap in per-capita gross domestic product in the Basque Country emerged over a two-decade period as a result of violence the Basque separatist group Euskadi Ta Askatasuna (ETA) utilized against the Spanish state.[115] There is also empirical evidence indicating that poor economic conditions enable resistance organizations to recruit better educated and better quali-fied personnel. In another article, Benmelech, Berrebi, and Klor found that a one-standard-deviation increase in the unemployment rate in the West Bank and Gaza Strip led to a 34-percent increase in the prob-ability that a suicide bomber had some academic education and a sim-ilar increase in the probability that a suicide bomber was previously involved in violent activities.[116]

Therefore, we see that empirical evidence supports the proposition that a government crackdown may promote mobilization in support of a violent resistance movement (at least among those directly impacted by violence), which may lead to a host population's higher tolerance for violence. Can one therefore claim that a resistance movement can gen-erally employ a strategy of provocation to boost its popularity and raise the upper threshold of violence? Before answering this question, we note that an incumbent government commonly wages a policy of repres-sion and collective punishment. In an analysis of thirty insurgencies

---

[p] To determine the magnitude of the estimated changes in the unemployment rate, the authors divided the value of the change for the district of origin by the average district-level unemployment rate. Thus, in the case of the change in the unemployment rate after a successful attack, the authors found that the unemployment rate increased by 0.52 percentage points, and they divided this estimate by the average unemployment rate for all Palestinian districts, which was approximately 9.8 percent. The result of this divi-sion is 5.3 percent.[113]

that began and were resolved between 1978 and 2008, Paul, Clarke, and Grill noted that counterinsurgent forces used a policy of repression and collective punishment in twenty of the thirty cases.[117] However, repression and collective punishment have a long history. Kalyvas noted that in March 1944 German forces in occupied Greece issued a public announcement indicating that sabotage would be punished with the hanging of three residents of the closest village unless the perpetrators were caught within forty-eight hours or it was shown that villagers had actively discouraged sabotage operations.[118] The announcement concluded by noting: "Hence the duty of self-preservation of every Greek when learning about sabotage intentions is to warn immediately the closest military authority."[119]

From an incumbent government's perspective, the logic behind a policy of repression and collective punishment is nicely captured by statements Napoleon Bonaparte made to one of his commanders:[120]

> Burn some farms and some big villages in the Morbihan and begin to make some examples . . . it is only by making war terrible that the inhabitants themselves will rally against the brigands and will finally feel that their apathy is extremely costly to them.

The underlying assumption of such a strategy, as Kalyvas noted, is that targeted noncombatants are somehow associated with insurgents and will therefore force them to cease their attacks against the government or affective bonds with noncombatants may cause insurgents to cease their attacks to spare noncombatants from further damage at the hands of counterinsurgents.[121] One mistake that counterinsurgents may make, though, as noted by Kalyvas, is to overestimate the affective bonds between insurgents and noncombatants or the influence of the latter on the former, as suggested by the following example:[122]

> However, insurgents may also disregard civilian demands, most likely when they come from villages with weak ties to them. The villagers of Malandreni, in the Argolid region of Greece, were told in April 1944 that a German officer would visit them on a set date. Upon learning of this visit, the Communist-led partisans decided to set up an ambush. Fearing German reprisals, the villagers demanded that the local Communist Party branch intervene with the partisans and have them cancel the ambush. The village party secretary describes the reaction of his regional boss: "Who do you think you are, comrade?" he was told; "A representative of the Germans?" To which he replied:

"No, comrade, I just came to compare the benefit [of ambushing the Germans] with its cost, this is why I came." "The Germans burned many other villages," the boss replied, "but these villages joined the partisans."

Rousing noncombatants from their apathy also appeared to be the Israeli strategy during Operation Cast Lead against Hamas in late 2008 and early 2009. As noted in one *New York Times* article written during the conflict:[123]

> The Israeli theory of what it tried to do here is summed up in a Hebrew phrase heard across Israel and throughout the military in the past weeks: "baal habayit hishtageya," or "the boss has lost it." It evokes the image of a madman who cannot be controlled.

> "This phrase means that if our civilians are attacked by you, we are not going to respond in proportion but will use all means we have to cause you such damage that you will think twice in the future," said Giora Eiland, a former national security advisor.

The article went on to note that Israeli actions led to Palestinians' deep rage toward Israel, but there were also some indications that residents of Gaza attempted to rein in Hamas. These two competing effects of Israel's policy of collective punishment suggest that Israeli decision makers, who are likely well versed in the conflict with the Palestinians and its nuances given the conflict's lengthy history, formulated subjective probabilities favoring the conclusion that the benefits accruing from Palestinian noncombatants reining in Hamas would exceed any negative effects that would follow from increased support and mobilization favoring Hamas.

Now we return to our earlier question of whether a provocation strategy is likely to be successful for a violent resistance movement. In the thirty insurgencies examined by Paul, Clarke, and Grill, twenty-two were won by insurgents, while the remaining eight were won by counterinsurgents. In eighteen of the twenty-two cases of insurgent victories, counterinsurgents attempted a strategy of repression and collective punishment, while in the eight cases of counterinsurgent victories, repression and collective punishment were used in only two cases.[124] Interestingly, in fourteen of the eighteen cases in which repression and collective punishment were used unsuccessfully, such tactics allowed counterinsurgents to win intermediate phases of the conflict on the way to defeat.[125] These results suggest that there is strong evidence to conclude that counterinsurgents should not employ a policy of collective punishment and repression. The authors noted:[126,q]

> Repression can win phases by dealing insurgents a blow and making support for them more costly, but our data show that the vast majority of phases that were won with repression ultimately increased popular support for the insurgency and ended in a COIN (counterinsurgency) defeat for the entire case.

Hence, there appears to be strong evidence supporting the notion that insurgents can use a provocation strategy to build support and potentially increase a population's tolerance for increasing levels of violence. Kalyvas also noted a number of cases in which an indiscriminate response backfired on the counterinsurgents. Regarding the use of indiscriminate violence by Germany during World War II, he noted:[128]

> The most infamous example of the futility of indiscriminate violence is possibly the Nazi reprisal policy in occupied Europe, aimed at deterring resistance against occupation. Reprisals appear to have been an utter and complete failure: they simply did not stifle resistance activity and, more importantly, they appear to have actually induced people to join the resistance. "Whatever the purpose of the German policy of reprisals," Condit points out, "it did little to pacify Greece, fight communism, or control the population. In general, the result was just the opposite. Burning villages left many male inhabitants with little place to turn except guerrilla bands. Killing women, children, and old men fed the growing hatred of the Germans and

---

q   The flip side of this analysis is the finding that resistance movements that (intentionally) target civilians typically fail to achieve their objectives. In an analysis of the target selections of all of the fifty-four groups that have ever been classified by the US Department of State as a foreign terrorist organization (FTO), Abrahms found that those groups that primarily targeted military rather than civilian targets were far more likely to achieve their political aims than groups that primarily targeted civilians. In particular, Abrahms categorized 125 kinetic campaigns carried out by these fifty-four groups as either "guerrilla" campaigns that primarily targeted military forces or "terrorist" campaigns that primarily targeted civilians. Of the 125 campaigns, thirty-eight achieved at least a partial success in terms of coercing a government into complying with a policy demand, and of this total, thirty-six were "guerrilla" campaigns that primarily targeted the military forces of a targeted state. The impact of target selection (i.e., civilian versus military) on outcomes was also found to be statistically significant in logistic regressions where the impact of other factors that could explain campaign outcomes, such as the strength and capabilities of the FTO and the targeted country, and the objective pursued by the FTO (either limited goals that did not seek to directly threaten a government or the way of life of its citizens or "maximalist" goals that did), were kept constant.[127]

the desire for vengeance." German observers in neighboring Yugoslavia "frankly concluded that rather than deterring resistance, reprisal policy was driving hitherto peaceful and politically indifferent Serbs into the arms of the partisans." Nazi reprisals produced a similar effect all over occupied Europe.

In addition to German conduct during World War II, Kalyvas also cited a number of other interesting examples:[129]

> Writing about the Vendeé War in 1797, Grachus Babeuf observed that the violent measures of the Republicans against the Vendean insurgents "were used without discrimination and produced an effect that was completely opposite to what was expected." A Greek guerrilla leader in Ottoman Macedonia at the start of the twentieth century asserted that a judicious balance had to be used in the administration of violence "for indiscriminate killing does harm rather than good and makes more enemies"; another one remarked that "the art is to find who should be punished." . . . Henriksen affirms that in "revolutionary warfare, reprisals serve the rebels' cause." He notes that in colonial Mozambique, "again and again, FRELIMO converts pointed to Portuguese acts as *the* prime factor for their decision. Non-Portuguese observers substantiated this assertion."

Hence, there appears to be strong empirical evidence suggesting that a resistance movement can use violence to promote mobilization and potentially raise the upper threshold of violence (if it succeeds in goading an opposing force into using a level of indiscriminate violence that inflames a population but which does not incapacitate a resistance). These findings suggest that violence, when used by counterinsurgents, may only be "effective when selective," that is, when used solely against insurgents and those individuals who actually provide material support, such as supplies, information, and sanctuary, to a resistance movement.[130]

However, various authors raised several qualifications to these observations. Wood noted that the argument that a population traumatized by an indiscriminate counterinsurgency campaign would flock to the insurgents rests on the overemphasized assumption that the insurgents have the capacity to protect civilians and safeguard their livelihood, which may not be the case.

He stated:[131]

> While regime backlash may mobilize civilians who
> were already at or close to the point of indifference
> between remaining neutral and supporting the reb-
> els, it does not necessarily follow that most civilians
> would support insurgents in the wake of indiscrimi-
> nate regime violence. Without credible security guar-
> antees from the rebels, civilians likely have insufficient
> incentive for supporting the risk of insurgents. Indeed,
> civilians may blame the rebels for the escalation in vio-
> lence and withhold support. Even when government
> forces kill large numbers of civilians, destroy property,
> and use other forms of collective punishment, civilians
> choose to collaborate with the incumbent's forces if
> the rebels are seen as weak.

As an example, Wood cited the case of the Tigray People's Liberation Front (TPLF), an Ethiopian group that fought against a ruling
military committee known as the Derg, which came to power after the
overthrow of the Haile Selassie regime in 1974. Wood noted:

> The TPLF's experience in Afar is telling. The Derg's
> campaign of repression against the Afar generated
> significant grievances among the population but was
> insufficient to drive the people into the arms of the
> rebels. In order for the rebels to profit from Mengis-
> tu's violence, they had to credibly demonstrate their
> commitment to making positive contributions to the
> lives of the Afar. This meant providing benefits such
> as economic development, political and educational
> structures, security, and justice systems.[132]

Frustration with the inability of insurgents to provide public goods,
including security, was also evident in Darfur, as indicated by the comments a sheikh in the town of Labado in Darfur made about the Sudan
Liberation Army (SLA): "We are angry at the SLA because they cause
us this bad situation. All of our wealth and our homes are taken, but
they run away and don't defend us."[133] Another example is the remarks
denouncing Hamas from the Palestinian woman whose house was
destroyed by Israel, as discussed earlier in this work.

Kalyvas also cited a number of other historical examples, arguing that a counterinsurgent force need not invest the resources to carry out a more selective and costly[r] campaign of violence and instead can rely on a more cost-effective strategy of indiscriminate violence when insurgents are weak:[135]

> This analysis yields the following prediction: incumbents can afford to be indifferent about the type of violence they use when insurgents are unable to offer any protection to civilians. Put otherwise, costly discrimination can be dispensed with when insurgents are weak. When this is the case, indiscriminate violence does succeed in paralyzing an unprotected population. When American indiscriminate violence made the Filipino civilians "thoroughly sick of the war," they "were forced to commit themselves to one side"; soon garrison commanders "received civilian delegations who disclosed the location of guerrilla hideouts or denounced members of the infrastructure." Likewise, most Missourians turned to the Union in their despair, Fellman notes, "not out of a change of faith but as the only possible source of protection." Guatemala provides the paradigmatic case in this respect. After the Guatemalan army used massive indiscriminate violence against the population, civilians who had initially collaborated with the rebels were left with no choice but to defect, because the rebels utterly failed to protect the population from the massacres. As Stoll points out, "while the guerrillas could not be defeated militarily, they were unable to protect their supporters."

Hence, these examples suggest that a provocation strategy may not succeed if a violent resistance movement is not capable of providing security against an indiscriminate government response. As suggested in the cited examples, when non-elites carefully weigh the costs and benefits of different actions and potential end states, ideological preferences (e.g., communism, a "liberated" Jerusalem) may give way to more immediate concerns associated with well-being and security, particularly for those who do not form the hard-core element of a movement.

In addition to whether an insurgent force is capable of providing security, there are other factors that need to be considered when

---

[r] Carrying out a campaign of selective violence is more costly, as Kalyvas noted, because it requires a complex and costly infrastructure to identify, locate, and neutralize insurgents.[134]

determining whether a provocation strategy can be used to increase the threshold of violence. First, there may be certain "structural" conditions that favor an indiscriminate strategy, thereby diminishing the utility of a provocation strategy. For instance, Downes argued that indiscriminate violence may be effective when used against a relatively small population in a confined geographic space because it permits counterinsurgents to imprison or kill the entire population, thereby effectively interdicting the provision of supplies, recruits, and information to insurgents.[136] As an example, Downes cited British tactics during the Second Anglo-Boer War of 1899–1902, which consisted of indiscriminate farm burning combined with the confining of the entire Boer civilian population of more than 110,000 into thirty-four squalid concentration camps.[137] The farm burning destroyed the food supply, while the effort to control the population through internment precluded any possibility that the population could assist the insurgents. The internment led to the death of more than 46,000 Boer and African civilians, many of whom died of preventable diseases caused by malnutrition, overcrowding, and unsanitary conditions.[138] Additionally, a confined geographic space was a factor, as the territory of the two Boer republics was already small before Britain adopted the scorched-earth policy.[139] This made it easier for the British to clear the veld of the entire Boer and African population, and the combination of these tactics led to the defeat of the insurgents.[140,s]

Another factor that needs to be considered, as noted earlier, is the degree of ties and affinity between insurgents and host populations. This was discussed previously in the context of collective punishment and from the perspective of insurgents and their interest (or lack thereof) in the welfare of noncombatants. However, it is also relevant from the perspective of noncombatants and their attitudes toward insurgents because the decisions of the former may materially impact the welfare and prospects of the latter. An interesting example is the reaction of the Shia in southern Lebanon after Israel invaded in 1982 to root out the presence of the Palestine Liberation Organization (PLO). An innocent observer might expect the Shia to side with their Muslim coreligionists in the PLO. However, the Shia population was sick of the depredations of the PLO and blamed the group for Israeli reprisal attacks before the 1982 invasion. In fact, the Shia in southern Lebanon initially welcomed the Israeli invasion, greeting incoming troops

---

s Another example cited by Downes included Italy's suppression of the Sanusi-led insurgency in the Cyrenaica region of Libya during 1923–1932. In 1930, the Italians attempted to "drain the sea" by interning the entire population of Cyrenaica and severing the Sanusis' supply lines from Egypt. Of the 85,000–100,000 individuals who were confined to camps, only 35,000 survived the war, and the relatively small size of Cyrenaica and of the population facilitated the Italian victory.[141]

with handfuls of thrown rice.[142] Opinion would soon turn against the Israelis when it became apparent they planned a long-term presence in the region. Nonetheless, the larger point is that by the time of the 1982 invasion, ties between the local Shia population and the PLO were strained, which impacted the population's level of tolerance for PLO violence against Israel. This example indicates that it is necessary for leaders of violent resistance movements to have a good read of the distribution of attitudes among a population when considering whether to use violence that may bring reprisals.

The fact that Shia attitudes were relevant in the contest between the PLO and Israel in southern Lebanon points to the larger issue of the role of civilian agency (i.e., the ability of a civilian populace to affect outcomes) in conflicts between an incumbent force (domestic or foreign) and a resistance movement. Indeed, civilian agency is *the* driving force underpinning the threshold of violence conceptual model, and Condra and Shapiro found empirical evidence that it played a role in combat against insurgent forces in Iraq. Analyzing significant activity reports of coalition forces in Iraq between February 2004 and February 2009, the authors argued that the Iraqi population punished the actor it deemed responsible for collateral damage.[143] Specifically, if coalition forces were blamed for collateral damage, insurgent attacks on coalition forces increased in subsequent periods. The authors attributed this increase to a decrease in the sharing of information about insurgents by noncombatants with coalition forces (which in turn allowed the latter to target insurgents and disrupt operations, thus reducing the ability of insurgents to produce violence). This in turn implies that (in this case) collateral damage caused by coalition forces raised a population's upper threshold of violence because the withholding of such information led to an increase in insurgent attacks on coalition forces in Iraq. In contrast, collateral damage attributed to insurgents resulted in fewer subsequent attacks, which the authors attributed to greater information sharing between host populations and coalition forces.[144, t, u]

Another tactic that violent resistance movements can use to affect the upper threshold of violence is the provision of economic aid and social welfare benefits to win "hearts and minds." Some groups hope this tactic will translate into greater host-population support for their militant activities. This logic is nicely captured by Bloom, who cited the case of Hamas:[149]

> Hamas spokespersons acknowledge that the group sees its sizeable social programs as a means of building and maintaining popular support for its political goals and program, including its militant and armed activities. "The political level is the face of Hamas, but without the other divisions Hamas would not be as strong as it is now," according to Ismail Abu Shanab. ". . . It needs the three parts to survive. If nobody supports these needy families, maybe nobody would think of martyrdom and the resistance of occupation." Another Hamas leader, Ibrahim al-Yazuri, characterized Hamas's objective as "the liberation of all Palestine from the tyrannical Israeli occupation . . . which is the main part of its concern. Social work is carried out in support of this aim."

Hence, for Hamas, the political intent of economic aid is to buttress the will of the people to support violent resistance against Israeli

---

t In particular, the authors found that in mixed areas where neither Sunnis, Shia, nor Kurds individually constituted more than 66 percent of the population, a one–standard-deviation increase in the number of civilian casualties caused by insurgents led to approximately 0.5 few attacks in the following week, which represented a 12-percent drop in the average number of attacks per 100,000 people in a mixed area.[145]

u Condra and Shapiro also uncovered temporal effects: the increase in attacks against coalition forces lasted for several weeks, after which violence against the coalition returned to its previous trend.[146] These findings are similar to those of Jaeger et al., who found that the relative support for radical Palestinian factions increased only for a few months after an increase in Palestinian fatalities caused by Israel.[147] These results suggest that any boost in the threshold of violence from a single violent event attributed to counterinsurgents is temporary and decays over time. A related issue is whether a population becomes desensitized to repeated acts of violence conducted by either or both actors. Atkinson and Kress were unable to find any studies within the behavioral psychology literature that examine how individuals process, react, and remember multiple violent or traumatic events. Instead, in developing their mathematical model of counterinsurgency, they relied on notions of "primacy" and "recency" found in the behavioral psychology literature: primacy suggests that the first experience of an event shapes subsequent behavior, while recency postulates that the latest event is more influential. In modeling a population's reaction to violence, this factor was implemented as a parameter that reflected either primacy or recency.[148]

control over the West Bank and Gaza Strip and indeed against Israel's existence. This strategy parallels Hizbollah's efforts to establish a "society of resistance" that is ever ready for war with Israel.[150] As part of this effort, Hizbollah provides a number of Iranian-funded social welfare services for fighters and party members and for Lebanese affected by the conflict with Israel. For instance, the Martyrs Foundation, an Iranian organization, made an important contribution to the provision of health care services in Lebanon, principally with the construction of the al-Rasul al-Azam Hospital in Dahiya, a southern suburb of Beirut.[151] All medical expenses of injured Hizbollah fighters are paid for at this facility, while 70 percent of expenses are covered for injured civilians. Interestingly, during elections, Hizbollah volunteers transport patients and staff at this facility to and from the polls.[152] The foundation also established, in Beirut and in Lebanon's Bekaa Valley, vocational schools for the daughters of fallen Hizbollah fighters and funds subsidized workshops to employ them.[153]

Another important social welfare benefit financed by Iran and provided by Hizbollah to Lebanese affected by combat is the reconstruction of homes. In particular, Jihad al-Bina' (Holy Reconstruction Organ), a construction company run by Hizbollah, partners with the Martyrs Foundation to rebuild homes destroyed during combat with Israel. Reportedly, one month after the 1996 Israeli military campaign Grapes of Wrath, Jihad al-Bina' rehabilitated more than 2,800 structures damaged by Israel in 106 locations in south Lebanon,[154] and overall between 1993 and 2006, the organization is estimated to have rebuilt nearly 15,000 homes.[155] Hizbollah collaborates with the Martyrs Foundation on reconstruction, with the insurgent group determining the validity of families' housing needs and, if necessary, arranging for the required property transactions. Financing provided by the Martyrs Foundation is used to acquire land, for which Jihad al-Bina' subsequently develops plans and builds the finalized structures.[156] Recent press reports indicate that Iran has spent $400 million to rebuild Dahiya after the July 2006 war with Israel.[157]

Another notable contribution to social welfare made possible by funding from the Martyrs Foundation is the construction by Jihad al-Bina' of 4,000-liter water reservoirs in each district of Beirut's heavily Shia southern suburbs.[158] Each reservoir was filled five times a day from continuously circulating tanker trucks, and in the absence of state-provided electricity in this area until 1990, generators mounted on trucks went to different buildings to provide the electricity required to pump water from private cisterns. As of 2006, Jihad al-Bina' served as the main source of drinking water for 500,000 people.[159] Jihad al-Bina' also provides public refuse collection for the half-million residents of Dahiya.[160]

In addition to the provision of pubic goods and social welfare benefits, groups can also potentially employ narratives to raise the upper threshold of violence.[v] Halverson, Goodall, and Corman defined a narrative as a "system of stories" or a collection of stories that relate to one another through coherent and consistent themes.[162] Additionally, Bruner noted that narratives often have universal qualities despite the particular circumstances under which they arose. One of these qualities is the centrality of trouble. Bruner noted that "stories worth telling and worth construing are typically born in trouble."[163]

Seminal works by a variety of cognition scholars have long argued that people package information gathered from everyday life into narrative form.[164] A narrative structure is thought to be critical for individuals to make sense of information and assign it meaning. When confronted with the complexity of actors, motivations, goals, and underlying cause-and-effect relationships, narratives are especially important tools for turning disparate facts into comprehensible stories and, as will be argued, potential drivers of violence.[165]

Hence, relevant for our purposes is the notion that violent resistance movements can potentially use narratives to encourage people to engage in (violent) collective action against an incumbent government or to persuade a host population to accept the legitimacy of, or at a minimum passively accept, a resistance movement's use of violence. Increased legitimacy in turn raises (or at least maintains) a population's tolerance for a group's use of violence.

How can violent resistance movements use narratives to potentially affect the threshold of violence? One way to approach this question is to use social movement theory (SMT) to deconstruct insurgent narratives and understand how they promote collective action. Sociologists originally developed SMT to understand the formation and evolution of a variety of movements, such as the civil rights and the pro-life movements in the United States. Although this theory was initially applied to (mostly) nonviolent social movements, the tools and framework of SMT can also be applied to understanding the narratives of insurgent movements.

A fundamental concept within SMT is that of a "frame." Erving Goffman first defined frames as "schemata of interpretation" that enable people "to locate, perceive, identify and label" events they experience or that are brought to their attention.[166] More simply, a frame represents a worldview or paradigm through which events and concepts are

---

[v] Portions of this section have been adapted from Agan's study on narratives.[161]

interpreted, and thus it represents a means through which meaning is constructed and "reality" is interpreted. Within the context of a social movement, by assigning motives and meanings, a frame can help overcome the collective action problem that is inherent in many insurgencies. As such, Benford and Snow define a "collective action frame" as "*action oriented* [italics added] sets of beliefs and meanings that inspire and legitimate the activities and campaigns of a social movement organization."[167] Thus, collective action frames perform an interpretive function by simplifying and condensing the "world out there,"[168] especially in ways "intended to mobilize potential adherents and constituents, to garner bystander support, and to demobilize antagonists."[169] Therefore, an important question for our purposes is how a collective action frame can be used to raise the threshold of violence.

Benford and Snow noted that collective action frames perform three core framing tasks, specifically diagnostic framing, prognostic framing, and motivational framing.[170] The first identifies the problem and its victim and attributes the problem to responsible actors and causes. The second specifies a proposed solution and a strategy for carrying out corrective action. The third provides a rationale for engaging in remedial collective action, including an appropriate vocabulary of motive.[171] This task is necessary to minimize "free riding" and encourage collective action.

For insurgent groups, many diagnostic frames are essentially "injustice frames" that identify the population from which the group emerged as a victim of what the group regards as a historical injustice perpetuated by some other actor. For instance, the diagnostic frame of the PLO encompassed a narrative that emphasized the loss of Palestinian land to Jewish settlers; the displacement of much of the original Palestinian population to the West Bank, Gaza Strip, and surrounding Arab countries; and the establishment of a Jewish state on most of British-Mandate Palestine. In the case of the Shining Path, the diagnostic frame centered on the arrival of the Spanish and the downfall of the Inca Empire in the sixteenth century as the cause of the current economic and social misfortune of the indigenous population of Peru.

Hence, we can see that sometimes a diagnostic frame entails a historical "original sin" that (from the perspective of leaders of a violent resistance movement) needs redress, and often a prognostic frame will call for the use of violence to sweep away existing conditions to pave the way for the establishment of new political, social, economic, and diplomatic arrangements.

An example is provided in the form of text from Hizbollah's 1985 "Open Letter" addressing the group's attitudes and perceptions of Israel:[172]

> We see in Israel the vanguard of the United States in our Islamic world. It is the hated enemy that must be fought until the hated ones get what they deserve. This enemy is the greatest danger to our future generations and to the destiny of our lands, particularly as it glorifies the ideas of settlement and expansion, initiated in Palestine, and yearning outward to the extension of the Great Israel, from the Euphrates to the Nile.
>
> Our primary assumption in our fight against Israel states that the Zionist entity is aggressive from its inception, and built on lands wrested from their owners, at the expense of the rights of the Muslim people. Therefore our struggle will end only when this entity is obliterated. We recognize no treaty with it, no cease fire, and no peace agreements, whether separate or consolidated.

In this passage, Israel figures prominently in Hizbollah's diagnostic frame as it represents "the greatest danger to our future generations and to the destiny of our lands" and "is aggressive from its inception, and built on lands wrested from their owners, at the expense of the rights of the Muslim people." The prognostic frame, which specifies what needs to be done, is fairly clear in its call for the use of violence: "Therefore our struggle will end only when this entity is obliterated."

Undoubtedly, the members and leadership of Hizbollah believe passionately in the themes represented within the diagnostic and prognostic frames captured in this section of the letter. However, narratives such as these also serve an instrumental purpose in terms of assisting with recruitment and, more generally, generating greater acceptance within a host population for a group's activities, including violent actions. More specifically, a group's narrative can potentially bolster legitimacy for its activities if it exhibits high fidelity with what are known as "master narratives." Halverson, Goodall, and Corman defined a master narrative as a "transhistorical narrative that is deeply embedded in a particular culture."[173]

Such narratives embody themes that deeply resonate with a population or culture, and Bernardi et al. noted that master narratives frequently appear over time and play a hegemonic role within a culture's narrative landscape:[174]

> They [master narratives] exercise mastery over competing narratives in the culture's narrative terrain not by eliminating but rather by marginalizing these alternative narratives. This repetition suggests a high frequency of dominant encoding on behalf of authors and decoding on behalf of readers. This frequency serves to cement these narratives as central components of a community's sense of history and perspective on the world.

Bernardi et al. also noted that narratives provide an explanatory framework and are often derived from foundational religious texts, such as the Koran, Bible, and Torah.[175] They may also draw from historical experience; the authors noted that Islamist extremists tended to draw from master narratives that invoke the Crusades or European imperialism in the Middle East during the modern era.[176] One recent example of this phenomenon is a video released by al-Qaeda in the Islamic Maghreb justifying its use of violence by linking it with resistance to the commencement of the French occupation of Algeria in 1803:[177]

> The war that is being waged by the jihadist group in the lands of the Islamic Maghreb is a legitimate war that seeks to defend against the diverse Crusader attack against Islam and its people in our Maghreb countries. This began with the wicked French occupation of Algeria in 1803 and is still continuing until this very day.

Another master narrative resonant within the Islamic world is that of the *Nakba*,[178] translated as "the catastrophe" in Arabic, which is the term used by Arabs to describe the establishment of the state of Israel in May 1948. In an extensive attack on Ayman al-Zawahiri, then second in command of al-Qaeda, published in the Egyptian daily *Al-Masry Al-Youm*, Sayyid Imam al-Sharif, a former associate of Al-Zawahiri who has since renounced violent jihad, criticized al-Qaeda for using the Palestinian cause as an empty talking point:[179]

> It is well-known that the fastest way to gain popularity among the Arab and Muslim masses is to bash the United States and Israel and talk a great deal about the Palestinian issue. Nasser did it, Saddam did it,

> Ahmadinejad does it, as do others. However, these peo-
> ple have actually done something for Palestinians . . .
> whereas Bin Laden and Z [Al-Zawahiri] just talk.

While the phrase *just talk* is used in a pejorative sense in this statement, the use of narratives to enshroud violent actions that are actually
taken can potentially infuse actions with legitimacy and raise a population's threshold of violence.

Hizbollah's "Open Letter" taps into each of the themes captured
by the master narratives previously cited. Israel is generally seen within
the Islamic world as a neocolonialist implant and usurper of sacred
Muslim territory that illegitimately wrested control of the land of Palestine from native Palestinians. Karagiannis noted that Hizbollah has
traditionally proffered a "Jerusalem liberation frame" in its narrative,
in which the group frames its military actions against Israel as a religious duty for devout Muslims to "liberate" Palestine and Jerusalem
from infidels.[180] By traditionally adopting such a frame in its narrative,
the group has tried to build greater support within the broader Arab
and Muslim world for its activities, including the use of violence.[w]

Hence, one way in which a group can attempt to influence the
threshold of violence is to enshroud its violent activities in narratives
that are highly resonant with the master narratives of a host population. A similar strategy is to equate the use of violence today with the
heroic and legitimate activities of canonical figures from the past. Such
was the rhetorical strategy of the Shia revolutionaries who successfully
overthrew the Shah in the late 1970s. The revolutionaries sanctified
violence and self-sacrifice in the name of the Islamic revolution by jux-
taposing extant violence with the exemplary martyrdom of canonical
Shia imams at the time of the sect's founding. They also referenced
other religious figures, thereby implying that current violence would
be rewarded with personal and eternal salvation in the afterlife, along
with a privileged position next to the pantheon of martyred Shia imams.

For instance, in one Friday prayer sermon delivered in March 1980,
Hojjat al-Islam Sayyid Ali Khamenei, the current supreme leader of
Iran, stated that "the martyrs are addressing us, [saying] 'you must
continue our way,'" and that they "want us to guard the heritage of
their blood, that is, the Islamic Republic and the victory of the Islamic
Revolution."[182] Additionally, the martyrs are "encouraging us" to hold
"the Koran in one hand and a gun" in the other, to "wage war against
all plots and cowardliness."[183] Additionally, in a sermon delivered in

---

[w]   It serves to recall the comment made by the individual with close contacts with
Hezbollah who noted that "Israel crossed a red line, and if Hezbollah did not react, Israel
will not stop . . . [the attack] shows that Hezbollah's confrontation is with Israel, so it can
get back its respected position in the Arab world."[181]

October 1979, Ayatollah Hussein-Ali Montazeri, commemorating Iranians killed in combat against Kurds in the Kurdish region of Iran, noted that the former were "martyrs" who "drank the elixir of martyrdom." As such, they were "gathered with the martyrs of the advent of Islam."[184]

Iran's revolutionary leaders also sought to depict the Islamic revolution as a continuation of the glorious struggles against tyranny waged by past prophets. One sermon from May 1984 noted:

> Moses revolted against the Pharaohs; Abraham against Nimrod and the Nimrods; Jesus against the tyrants . . . of his time; and Mohammad . . . too, delivered the people from the hands of oppressive rulers. In contrast to what some self-interested people say— that the prophets had always been instruments . . . of oppressors . . . they always . . . stood up against the taghuts [tyrants, idolaters].[185]

Additionally, in 1980, Khamenei noted that "our Islamic Revolution . . . was, therefore, a resumption of the revolution of the prophets, [because it] stood up against injustice from its inception . . . [against] the . . . oppressors who rules [sic] Iran."[186]

In this case, equating current uses of violence with the heroic acts of exemplary figures from the past essentially constitutes a motivational frame because it encourages participation (rather than free riding) by implying infinite rewards in the afterlife. Even beyond attracting new converts willing to commit violence, however, such appeals to a glorious tradition represent a group's effort to legitimate its uses of violence in the eyes of the broader population, particularly if the tradition resonates with a host population.

Narratives that attempt to draw a connection between current violence and the activities of past canonical figures, or that resonate with extant master narratives within a society, may represent efforts to increase or at least maintain a population's threshold of violence. Is there any empirical evidence that narratives, and more generally discourse, can be used in either of these manners to successfully raise a population's threshold of violence? Bernardi et al. noted that shortly after a 2005 bovine inoculation campaign instituted by multinational forces in Iraq, a rumor spread that US forces had begun an effort to starve the Iraqi population by poisoning its livestock.[187] Although it is unclear whether the rumor was spread by Iraqi insurgents or arose organically, it proved potent at that time because not only was Iraq undergoing an existential and sociopolitical crisis, but also a large amount of its livestock was dying because of disease and a significant water shortage.

The conspiratorial nature of the rumor found a receptive audience, given Iraqi frustration with political, social, and economic conditions in post-Hussein Iraq and Iraqi suspicions that the United States invaded Iraq to pillage its oil resources. Bernardi et al. noted that the rumor provided a convenient target for pent-up frustrations, anxieties, and fears, which led farmers to turn a blind eye to insurgent activities and even to participate in violence.[188] Hence, the rumor potentially contributed to an increase in the threshold of violence among a group of people, and even if it was not initially spread by insurgents, it represents an example of how a narrative could be used by insurgents to gain greater acceptance of violent activities.

The rumor alleging sinister American intentions circulated within an environment that featured master narratives related to the crusades and European colonialism, and the fidelity between the rumor and these master narratives made it more potent, as the authors noted:[189]

> The bovine poisoning mosaic implicitly invoked the Crusader master narrative in which non-Arab/non-Muslim invaders plunder Arab lands for their riches while proclaiming allegiance to a higher power and a more righteous religion. The wanton killing of livestock and the imposition of privation through drought in this mosaic parallel the slaughter of Muslims by Christian soldiers of the First Crusade. The accusation of economic exploitation parallels the nineteenth-century reworking of the Crusader narrative in the form of colonization . . . This master narrative of invasion, destruction, and exploitation eventually defined the US occupation across Iraq and, even more problematically, the Islamic world, undermining US strategic messaging and, as a result, US strategic planning and vision.

> More important, the association of the bovine rumor mosaic with this Crusader narrative enhanced its threat: rumors that activate, catalyze, and extend this historical and widely understood narrative about foreign invaders are both more problematic because of their implications and more widespread because of their narrative fidelity. In other words, because they follow the same pattern and underlying message as the Crusader narrative, these rumors will find widespread traction and belief because they reinforce a prevailing narrative of invasion and exploitation.

The authors called such rumors "narrative" improvised explosive devices (IEDs) because they help "entice contested populations into a kind of complacency, even tolerance, for kinetic IEDs targeting US and government forces."[190] Hence, the strategic use of narratives (and their association with relevant master narratives) represents another potential tool violent resistance movements can use to increase a population's threshold of violence.

CONCLUSION

In this report, we provided insights into the dynamics of the threshold of violence model and the potential tactics violent resistance movements use to increase a population's tolerance for a group's use of violence against an incumbent government. We identified several potential strategies: the use of narratives and the provision of public goods and social welfare benefits to promote a "society of resistance" and the use of a provocation strategy to goad an incumbent government into a disproportionate response.

The threshold of violence model posits a situation in which rational decision makers carefully evaluate the costs and benefits of using violence, attempt to anticipate their adversary's actions, and subsequently choose a course of action that maximizes the benefits or utility to their group. While it represents a potentially useful framework for understanding the decision to use violence, there are other perspectives through which to evaluate this decision. Thus, it may be useful to briefly discuss other paradigms that explore the decision to use violence and highlight various limitations and qualifications to the threshold of violence model.

First, identity, and the desire for positive group status at the expense of other groups, may influence cost–benefit calculations surrounding the use of violence. Social psychological experiments conducted by Billig and Tajfel in the 1970s showed that participants placed in different groups exhibited strong group loyalty even when their group assignment was based on random (i.e., through a coin toss) or trivial factors and even in the absence of face-to-face interactions with group and non-group members.[191] Further experiments indicated that subjects, when prompted to apportion fictitious rewards between groups, preferred outcomes that maximized intergroup differentials and relative gains even if this preference meant less absolute gains for one's group. Such findings led Horowitz to argue that the desire to enhance relative group status was the prime motivation for ethnic conflict.[192]

The importance of positive group status also plays an important role in Petersen's emotion-based theory of ethnic conflict. Petersen argued that emotions such as fear, hatred, resentment, and rage play an important role in determining the targets and timing of ethnic violence.[193] For instance, he noted that group-based resentment explains why Ukrainians in Soviet-occupied Poland attacked Poles once eastern Poland fell under Soviet control in September 1939 and why they attacked Jews instead of Poles in 1941 after the German invasion of Soviet territory.[194]

In the former case, land, language, and education policies adopted by the interwar Polish government led to an ethnic hierarchy in which Poles were privileged at the expense of Ukrainians, despite both groups each constituting about a third of the population in eastern Poland.[195] However, the Soviet occupation led to the devastation of the Polish community, with hundreds of thousands deported and many others victims of score-settling violence by Ukrainians. Additionally, Polish property was expropriated, Polish nationalist symbols were banned, and the language of instruction in educational institutions throughout eastern Poland switched from Polish to Ukrainian.[196]

The status of Jews in eastern Poland, which before September 1939 had been no higher than that of Ukrainians, was elevated by the Russian takeover, though. Jews, who had played an important role in the Communist Party in Poland, were recruited heavily into the new Soviet administration.[197] Jan Karski, a Pole who wrote reports for the Polish government in exile, observed:[198]

> They (Jews) are entering the political cells; in many of them they have taken over the most critical political-administrative positions. They play quite a large role in the factory unions, in higher education, and most of all in commerce; but, above and beyond all of this, they are involved in loansharking and profiteering, in illegal trade, contraband, foreign currency exchange, liquor, immoral interests, pimping, and procurement. In these territories, in the vast majority of cases, their situation is better both economically and politically than before the war.

Jews rather than Poles had become the Ukrainians' status rival under the Soviet occupation, so after the German invasion, the Ukrainians attacked the Jews to "put them back in their place."[199] As this analysis indicates, calculations surrounding the threshold of violence may become clouded by group status concerns and various violence-inducing emotions that may be widespread among a population.

More broadly, Gurr noted that psychological frustration in the form of "relative deprivation" is a fundamental precondition for civil strife.[200]

Gurr defined relative deprivation as the perception of the discrepancy between "value expectations" (i.e., the goods and conditions of life to which individuals believe they are justifiably entitled) and "value capabilities" (i.e., the ability to attain and keep value expectations). Violence and, more broadly, civil strife are more likely the greater the magnitude and scope of relative deprivation within a population. Furthermore, as seen in the interactions between Jews, Poles, and Ukrainians in the period immediately before and after the Nazi invasion of the Soviet Union, relative deprivation combined with group status concerns can potentially contribute to the identification of targets of violence.

Additionally, if a conflict between an insurgent movement and an incumbent government or outside force places at risk values deemed "sacred" by a host population, the latter may be more willing to tolerate violence to protect such values. In particular, Ginges et al. argued that, once certain issues become sacred, they may no longer be amenable to conventional cost–benefit calculations because individuals tend to resist (and regard as profane) attempts to buy off their commitments to such values.[201] For instance, in a poll of Israeli settlers, 46 percent of those surveyed indicated it was never permissible for Jews to forfeit parts of the "Land of Israel" to Palestinians as part of a peace deal, regardless of the value of any benefits received in return. Similarly, 54 percent of surveyed Palestinian students indicated it was never permissible to compromise on the "right of return" of Palestinian refugees or over sovereignty over Jerusalem. Furthermore, more than 80 percent of surveyed Palestinian refugees indicated their unwillingness to compromise over the right of return to their original homes, irrespective of the benefits that may accrue from a peace deal.[202]

Clearly, such results indicate that, for many people, sacred values are not fungible resources with a value that can be quantified and sacrifice that can be compensated through some form of material incentive or inducement. In fact, among such "moral absolutists" within the surveyed Jewish and Palestinian populations, their support for a violent opposition to a peace deal increased when they were asked to consider a hypothetical deal that involved some form of material compensation for a compromise on a sacred value.[203] Interestingly, opposition among moral absolutists decreased when a hypothetical deal entailed a symbolic compromise by the adversary on one of their own sacred values. For instance, support for a violent opposition to a compromise among Palestinian refugees decreased if an agreement required Israelis to relinquish what they believe to be their sacred right to the West Bank or if Israel symbolically recognized the legitimacy of the right of return.[204,x] Hence, sacred values seem to exist on a separate plane and are judged by a different calculus than non-sacred values; the survey results indicate that compromises on sacred values can be achieved if the other side makes equitable sacrifices on values they deem sacred. Nonetheless, the larger point for our purposes is that the presence of sacred values placed at risk may make a host population tolerate greater levels of violence (although such willingness may be tempered by the immediate material consequences of violence, as demonstrated by the previously noted Palestinian woman who blamed Hamas for the Israeli destruction of her home during Operation Cast Lead).

Furthermore, given its relevance to the threshold of violence model, we discussed at length the key features of the rational choice paradigm. However, it is also important to note other decision-making frameworks that may also be relevant in the decision by governments or insurgent leaders to use force. Developed in response to critiques that rational choice models do not accurately reflect actual decision making, prospect theory argues that how choices are framed for decision makers is an important factor in understanding decision making.[206] Framing effects are not captured in rational choice models, although they often manifest themselves within controlled experiments, as noted by McDermott:[207]

> The first experiment asked people to pretend that they were responsible for making public policy in the face of a major flu epidemic that was expected to kill 600 people. They were asked to decide between two different programs that were each designed to contain this epidemic. The choices were presented to the first group as follows: policy A will save 200 people; policy B has a one-third chance that 600 people will be saved, and a two-thirds chance that no one will be saved. In this case, 72 percent chose the *first* option. The second group was presented with these choices: policy A will cause 400 people to die; policy B has a one-third chance that no one will die; and a two-thirds chance that 600 people will die. In this case 78 percent chose the *second* option.

Both policies A and B entail the same number of expected survivals (200) and deaths (400), yet decision making differed drastically based on how the choice was framed: the first choice emphasized a "saving"

---

frame, whereas the second choice emphasized a "death" frame. Rational choice theory would predict that decision makers would be indifferent as to how a choice was framed, which was clearly not the case in this example.

Hence, prospect theory argues that how a choice is framed impacts decision making. Specifically, choices can be framed as gains (i.e., saving lives) or losses (i.e., deaths). Furthermore, if decision makers find themselves in a "domain of gains," or when things are going well, they tend to be risk averse, while in a "domain of losses," or when things are going poorly, they tend to be relatively risk seeking in an effort to recoup losses.[208] While prominent in the field of behavioral economics, prospect theory has also been used to evaluate political and military decision making. For instance, McDermott argued that prospect theory is useful in understanding President Jimmy Carter's April 1980 decision to attempt a military rescue of US diplomats held hostage by Iran. Specifically, she argued that because of domestic and international developments, Carter found himself in a domain of losses. Internationally, US prestige suffered in the wake of the hostage crisis, and Carter's failure to secure the release of the hostages during the first five months of the crisis led to a sharp drop in public support for the administration and increased Congressional pressure for the administration to resolve the crisis.[209] From the perspective of prospect theory, President Carter was operating in a domain of losses, which made him relatively more willing to accept risk when reviewing options because of his desire to reverse both domestic and international setbacks. Thus, prospect theory offers a potentially useful correction to the rational choice model and may therefore offer insights regarding the decision to use violence by insurgent leaders.

We can see that there are a number of potential qualifications to the threshold of violence model. Nonetheless, keeping these reservations in mind, in certain contexts the paradigm presented by the threshold of violence model can potentially be a useful starting point to understanding resistance movements' decisions to use violence. As we have seen, insurgents can potentially use a provocation strategy to raise a population's upper threshold of violence, and this strategy produces a number of competing effects. On the one hand, a government crackdown can promote insurgent mobilization and greater support for the use of violence if a counterinsurgent campaign produces widespread damage and personal loss throughout a population; on the other hand, the desire to limit damage and the lack of faith in the ability of insurgents to provide security may lead host populations to try to rein in the activities of violent resistance movements and even to seek the succor of the incumbent (assuming the host population does not conclude that the incumbent seeks its obliteration or displacement).[y]

These dynamics are represented in Figure 2, which shows a decision tree with the insurgent group as the initial decision maker deciding whether to initiate a provocation strategy. If a provocation strategy is employed, the incumbent, which can be a domestic authority or foreign power, then decides whether to initiate a discriminate or indiscriminate campaign, and the public decides whether to support the incumbent or the rebels. Perceptions of violence by the general public may be influenced by a number of qualifying factors, including whether counterinsurgent violence is perceived as proportionate (i.e., force is limited to what is necessary to achieve military objectives, and collateral damage is minimized), whether distinctions between military and civilian targets are observed, whether the use of violence against opposing forces is militarily necessary, and whether armed forces limit themselves to tactics and weapon systems that do not cause unnecessary injury or suffering.[z]

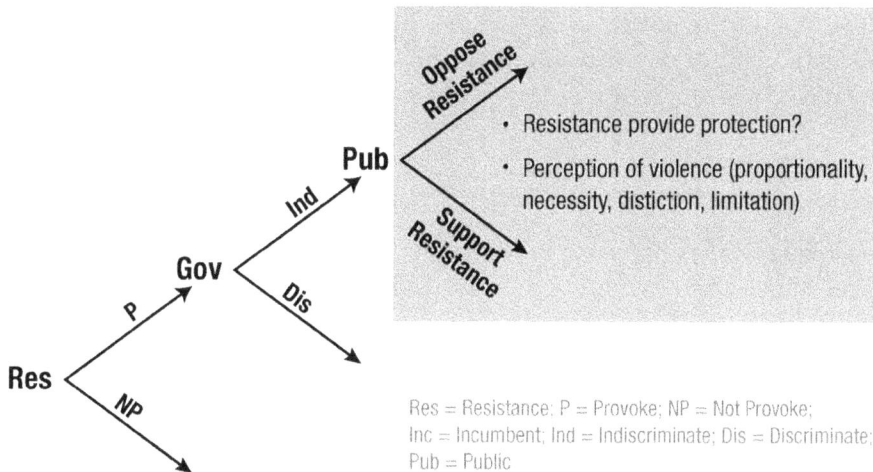

Res = Resistance; P = Provoke; NP = Not Provoke;
Inc = Incumbent; Ind = Indiscriminate; Dis = Discriminate;
Pub = Public

Figure 2. Provocation strategy decision tree.

Perceptions regarding the ability of a resistance movement to provide protection to a populace will likely factor heavily in whether

---

[y] For a mathematically rigorous treatment of these trade-offs and how they impact the decision making of violent resistance movements and host populations, see Berman et al., Bueno de Mesquita and Dickson, and Bueno de Mesquita.[210]

[z] These four factors (proportionality, necessity, distinction, and limitation) are the key principles underlying the law of armed conflict. For more on the law of armed conflict, see Law of War Manual.[211]

the latter sides with the incumbent or the resistance movement if the incumbent launches an indiscriminate campaign. As Kalyvas noted:

> Assume a setting where incumbents choose whether to use indiscriminate or selective violence, insurgents have the option of protecting civilians from incumbent indiscriminate violence, and civilians collaborate with the political actor who best guarantees their security. In such a setting, civilians will be likely to collaborate with the incumbents if the insurgents fail to protect them, whether incumbents are indiscriminate or selective; they will be likely to side with the insurgents when they are protected by them against indiscriminate incumbents; and the outcome is indeterminate when insurgents protect civilians and incumbents are selective.[212]

Lastly, we end our discussion by noting the paucity of theoretical models that can accurately depict the necessary and sufficient minimum level of violence a resistance movement must employ to secure a specific concession. As DeNardo noted, there is no formal model that can accurately indicate an optimal level of violence for any given situation. Although the comments below are in the context of social movements contemplating an escalation to violent tactics, they can also apply to violent insurgent and resistance movements contemplating how much violence to use to gain concessions from an incumbent or to implement a provocation strategy:[213]

> Given our reasoning about responsiveness and recruiting, the natural question to address at this point is whether any simple (monotonic or single-peaked) association exists between the level of violence used by the movement and the concessions it receives. Are tactics ranked in effectiveness, for example, strictly according to their proximity on the continuum of violence to the optimal choice, in such a way that the dissidents will converge directly on the best tactic by a simple process of trial and error? The answer to this question is that no simple relation necessarily exists between tactics and concessions . . . On the contrary, the optimal tactic can in general lie anywhere along the continuum of violence, and more importantly, there can be no assurance that other tactics will smoothly diminish in effectiveness as they diverge in either direction from the optimal choice. It follows that locating the optimal tactic will often be no easy matter for dissident

leaders and that a highly detailed assessment of the political environment (probably unattainable in most cases) will be necessary to predict in advance the relative potential of each kind of violence.

While it may be the case that a theoretical model of sufficient granularity and realism is currently out of reach for connecting violence with coerced concessions, empirical studies utilizing econometric and statistical techniques on micro-level data (e.g., data on specific tactics and weapon systems employed, geolocated data on casualties caused by insurgents and/or incumbents) may offer a fruitful path toward answering specific questions regarding whether certain tactics and operations violate a community's tolerance for violence. For instance, Benmelech, Berrebi, and Klor noted that a one-standard-deviation increase in *punitive* house demolitions by Israel (which targeted actual Palestinian suicide attackers and enablers) led to a *decrease* of 11.7 percent in the number of suicide attacks emanating from an average district.[214] Meanwhile, a standard deviation increase in the number of *precautionary* house demolitions, which were justified by the location of the house and were unrelated to the actions of the house owner, led to a 48.7 percent *increase* in the number of suicide attacks from an average district. This empirical finding bears directly upon the threshold of violence (because it identified a particular tactic that exceeded a community's tolerance for violence), and it affirms the earlier observation by Kalyvas that perceptions of fairness, in particular being subjected to violence independent of one's actions, impact a public's reaction to a counterinsurgent campaign.

Another empirical study with direct relevance to the threshold of violence was the analysis of collateral damage conducted by Condra and Shapiro using Iraqi data. As previously noted, the authors found that an increase in civilian casualties caused by coalition forces led to an increase in insurgent attacks on coalition forces, while an increase in civilian casualties caused by insurgents led to a decrease in such attacks. However, by decomposing insurgent attacks on coalition forces into various types, specifically direct fire, indirect fire, IED, and suicide attacks, the authors found that this trend was only present for direct fire insurgent attacks on coalition forces.[215] This form of attack, the authors noted, was the most susceptible to detection by noncombatants during the setup phase, and therefore, this study provides a more granular analysis of how violations of the threshold of violence might play out in practice.

As noted earlier in the text, the mechanism through which these dynamics are effectuated is through the provision (or non-provision) of information by noncombatants to coalition forces. Interestingly, and also as noted earlier, Condra, Felter, Iyengar, and Shapiro noted that this information effect was not present in Afghanistan. Rather, in the Afghan context, a revenge effect, whereby people directly impacted by coalition violence acted out revenge by joining a resistance movement offering such an opportunity, was the driving factor explaining increases in violence against coalition forces. The authors noted that an information effect is a short-run effect (because information can be shared and acted upon immediately), whereas a revenge effect involves a relatively lengthier process because it takes time for a resistance movement to incorporate new recruits.[216] Hence, if appropriate micro-level data exist, well-crafted empirical studies such as these may provide useful insights on mechanisms surrounding the practical unfolding of violations of the threshold of violence, and ensuing results can be compared across contexts.

NOTES

[1] Michael P. Atkinson and Moshe Kress, "On Popular Response to Violence during Insurgencies," *Operations Research Letters* 40, no 4 (2012): 223.

[2] Eric P. Wendt, "Strategic Counterinsurgency Modeling," *Special Warfare* 18, no. 2 (2005): 4–5.

[3] Maria J. Stephan and Erica Chenoweth, "Why Civil Resistance Works: The Strategic Logic of Nonviolent Conflict," *International Security* 33, no. 1 (2008): 7–44.

[4] Lionel Cliffe, Joshua Mpofu, and Barry Munslow, "Nationalist Politics in Zimbabwe: The 1980 Elections and Beyond," *Review of African Political Economy* 7, no. 18 (1980): 55.

[5] John Darby, "Legitimate Targets: A Control on Violence?," in *New Perspectives on the Northern Ireland Conflict*, ed. Adrian Guelke (Aldershot, England: Avebury, 1994), 63.

[6] Douglas Jehl and Thom Shanker, "The Struggle for Iraq: Terrorist Liaisons; Al Qaeda Tells Ally in Iraq to Strive for Global Goals," *New York Times*, October 7, 2005, http://www.nytimes.com/2005/10/07/world/the-struggle-for-iraq-terrorist-liaisons-al-qaeda-tells-ally-in-iraq-to-strive-for-global-goals.html?_r=0.

[7] John A. McCary, "The Anbar Awakening: An Alliance of Incentives," *Washington Quarterly* 32, no. 1 (2009): 44.

[8] Asoka Bandarage, *The Separatist Conflict in Sri Lanka: Terrorism, Ethnicity, Political Economy* (New York and Bloomington, IN: iUniverse, 2009), 154.

[9] Ibid.

[10] C. A. Chandraprema, *Sri Lanka: The Years of Terror. The JVP Insurrection 1987–89* (Colombo, Sri Lanka: Lake House Bookshop, 1991), 296.

[11] Wendt, "Strategic Counterinsurgency Modeling," 5.

[12] Gordon H. McCormick, "Terrorist Decision Making," *Annual Review of Political Science* 6, (2003): 481–486.

[13] Ibid., 481.

[14] Ibid., 482.

[15] Andrew H. Kydd and Barbara F. Walter, "The Strategies of Terrorism," *International Security* 31, no. 1 (2006): 63.

[16] Ami Pedahzur, *Suicide Terrorism* (Cambridge, UK: Polity Press, 2005), 44.

[17] Ibid., 55.

[18] Lawrence A. Kuznar, "Rationality Wars and the War on Terror: Explaining Terrorism and Social Unrest," *American Anthropologist* 109, no. 2 (2007): 320.

[19] Bryan D. Jones, "Bounded Rationality," *Annual Review of Political Science* 2 (1999): 297–321.

[20] Stephen M. Walt, "Rigor or Rigor Mortis? Rational Choice and Security Studies," *International Security* 23, no. 4 (1999): 5-48; Donald P. Green and Ian Shapiro, *Pathologies of Rational Choice Theory: A Critique of Applications in Political Science* (New Haven, CT: Yale University Press, 1994); for the applicability of rational choice theory to the study of terrorism, see Bryan Caplan, "Terrorism: The Relevance of the Rational Choice Model," *Public Choice* 128, no. 1–2 (2006): 91–107.

[21] Mick Moore, "Thoroughly Modern Revolutionaries: The JVP in Sri Lanka," *Modern Asian Studies* 27, no. 3 (July 1993): 637, 640–641.

[22] Ibid., 640–641.

[23] Ibid., 638

[24] Ibid.

[25] Caplan, "Terrorism: The Relevance of the Rational Choice Model," 93.

[26] Guillermo Pinczuk, Mike Deane, and Jesse Kirkpatrick, *Case Studies in Insurgency and Revolutionary Warfare—Sri Lanka (1976–2009)*, ed. Guillermo Pinczuk (Fort Bragg, NC: US Army Special Operations Command, 2014), 90.

[27] Bandarage, *Separatist Conflict in Sri Lanka*, 154.

[28] Diplomatic Correspondent, "Rajiv Assassination 'Deeply Regretted': LTTE," *The Hindu*, June 28, 2006, http://www.thehindu.com/todays-paper/rajiv-assassination-quotdeeply-regretted-ltte/article3125569.ece.

[29] "WarByAnyotherName," *The Economist*, July 5, 2014, 42–43, http://www.economist.com/news/europe/21606290-russia-has-effect-already-invaded-eastern-ukraine-question-how-west-will.

[30] Samuel L. Popkin, *The Rational Peasant: The Political Economy of Rural Society in Vietnam* (Berkeley: University of California Press, 1979), ix.

[31] T. David Mason, "Insurgency, Counterinsurgency, and the Rational Peasant," *Public Choice* 86, no. 1–2 (1996): 64.

[32] Ethan Bronner, "Parsing Gains of Gaza War," *New York Times*, January 19, 2009, http://www.nytimes.com/2009/01/19/world/middleeast/19assess.html?hp=&_r=0.

[33] McCormick, "Terrorist Decision Making," 484.

[34] James DeNardo, *Power in Numbers: The Political Strategy of Protest and Rebellion* (Princeton, NJ: Princeton University Press, 1985), 194.

[35] McCormick, "Terrorist Decision Making," 486–490.

[36] Ibid., 490–495.

[37] Martin C. Libicki, Peter Chalk, and Melanie Sisson, *Exploring Terrorist Targeting Preferences* (Santa Monica, CA: RAND Corporation, 2007), 19–20.

[38] McCormick, "Terrorist Decision Making," 487.

[39] C. J. M. Drake, "The Role of Ideology in Terrorists' Target Selection," *Terrorism and Political Violence* 10, no. 2 (1998): 60.

[40] Jakana Thomas, "Rewarding Bad Behavior: How Governments Respond to Terrorism in Civil War," *American Journal of Political Science* 58, no. 4 (2014): 804–818.

[41] Ibid., 806.

[42] Ibid., 813.

[43] Kydd and Walter, "The Strategies of Terrorism," 51.

[44] Ibid., 58.

[45] Geoffrey Blainey, *The Causes of War*, 3rd ed. (New York: Free Press, 1988), 122.

46 Per Baltzer Overgaard, "The Scale of Terrorist Attacks as a Signal of Resources," *Journal of Conflict Resolution* 38, no. 3 (1994): 452–478.

47 Kydd and Walter, "The Strategies of Terrorism," 50–51.

48 Ibid., 58.

49 Ibid.

50 Ibid., 59–60.

51 Ibid., 60.

52 Stephen Charles Nemeth, "A Rationalist Explanation of Terrorist Targeting" (PhD diss., University of Iowa, 2010), 34–36, http://ir.uiowa.edu/etd/718/.

53 Benedetta Berti, Institute for National Security Studies (Tel Aviv, Israel), quoted in Jodi Rudoren and Anne Barnard, "Hezbollah Kills Two Israeli Soldiers Near Lebanon," *New York Times*, January 28, 2015, http://www.nytimes.com/2015/01/29/world/middleeast/israel-lebanon-hezbollah-missile-attack.html?_r=0.

54 Rudoren and Barnard, "Hezbollah Kills Two Israeli Soldiers Near Lebanon."

55 Paul Staniland, "States, Insurgents, and Wartime Political Orders," *Perspectives on Politics* 10, no. 2 (2012): 243–264.

56 Rudoren and Barnard, "Hezbollah Kills Two Israeli Soldiers Near Lebanon."

57 "Back to Bashing: Messrs. Netanyahu and Nasrallah Flirt with War Once More," *The Economist*, January 31, 2015, http://www.economist.com/news/middle-east-and-africa/21641308-messrs-netanyahu-and-nasrallah-flirt-war-once-more-back-bashing.

58 Kydd and Walter, "The Strategies of Terrorism," 66.

59 Daniel Byman, "The Logic of Ethnic Terrorism," *Studies in Conflict & Terrorism* 21, no. 2 (1998): 160.

60 Ibid.; Thomas, "Rewarding Bad Behavior," 807.

61 Jeremy M. Weinstein, *Inside Rebellion: The Politics of Insurgent Violence* (Cambridge: Cambridge University Press, 2007).

62 Pinczuk, Deane, and Kirkpatrick, *Sri Lanka (1976–2009)*, 192.

63 Ibid., 308.

64 Byman, "The Logic of Ethnic Terrorism," 158.

65 Ibid.

66 Ibid.

67 Pinczuk, Deane, and Kirkpatrick, *Sri Lanka (1976–2009)*, 252.

68 McCormick, "Terrorist Decision Making," 484.

69 Kydd and Walter, "The Strategies of Terrorism," 69.

70 Ethan Bueno de Mesquita and Eric S. Dickson, "The Propaganda of the Deed: Terrorism, Counterterrorism, and Mobilization," *American Journal of Political Science* 51, no. 2 (2007): 365–366.

71 Stathis N. Kalyvas, *The Logic of Violence in Civil War* (Cambridge, UK: Cambridge University Press, 2006), 153–154.

72 Ibid., 154.

73 Ibid.

74 Ibid., 154–155.

75 Ibid., 151.

76 Ron Buikema and Matt Burger, "Sendero Luminoso (Shining Path)," in *Casebook on Insurgency and Revolutionary Warfare Volume II: 1962–2009*, ed. Chuck Crossett (Ft. Bragg, NC: US Army Special Operations Command, 2012), 96–97.

77 Mia Bloom, "Palestinian Suicide Bombing: Public Support, Market Share and Outbidding," *Political Science Quarterly* 119, no. 1 (Spring 2004): 68.

[78] Anthony York, "Israel Rally Critical of Bush," *Salon.com*, April 16, 2002, https://www.salon.com/2002/04/17/rally/.

[79] Bloom, "Palestinian Suicide Bombing," 68–73.

[80] Ibid., 69, 71, 73–75.

[81] Ibid., 69

[82] Ibid., 69, 73–74.

[83] Max Abrahms and Philip B. K. Potter, "Explaining Terrorism: Leadership Deficits and Militant Group Tactics," *International Organization* 69, no. 2 (2015): 311–342.

[84] David A. Jaeger, Esteban F. Klor, Sami H. Miaari, and M. Daniele Paserman, "The Struggle for Palestinian Hearts and Minds: Violence and Public Opinion in the Second Intifada," *Journal of Public Economics* 96, nos. 3–4 (2012): 355.

[85] David A. Jaeger, Esteban F. Klor, Sami H. Miaari, and M. Daniele Paserman, "Can Militants Use Violence to Win Public Support? Evidence from the Second Intifada," *Journal of Conflict Resolution* 59, no. 3 (2015): 529–530.

[86] Kydd and Walter, "The Strategies of Terrorism," 72–74.

[87] Ibid., 74.

[88] Summer Agan, *Narratives and Competing Messages* (Fort Bragg, NC: US Army Special Operations Command, forthcoming), 16–17.

[89] Mancur Olson, *The Logic of Collective Action: Public Goods and the Theory of Groups* (Cambridge, MA: Harvard University Press, 2009), 1–2.

[90] Gordon H. McCormick and Frank Giordano, "Things Come Together: Symbolic Violence and Guerrilla Mobilisation," *Third World Quarterly* 28, no. 2 (2007): 295–320.

[91] Ibid., 309.

[92] Ibid., 307; Thomas Thornton, "Terror as a Weapon of Political Agitation," in *Internal War*, ed. Harry Eckstein (Westport, CT: Greenwood Press, 1980), 74.

[93] Benedict Anderson, *Imagined Communities* (London: Verso, 2006), 5-6.

[94] Byman, "The Logic of Ethnic Terrorism," 154.

[95] Ibid.

[96] Ibid., 155.

[97] Chaim Kaufmann, "Possible and Impossible Solutions to Ethnic Civil Wars," *International Security* 20, no. 4 (1996): 144.

[98] Weinstein, *Inside Rebellion*; Paul Collier and Anke Hoeffler, "Greed and Grievance in Civil War," *Oxford Economic Papers* 56, no. 4 (2004): 563–595.

[99] Darby, "Legitimate Targets," 62–63.

[100] Ibid., 46.

[101] bid., 47.

[102] Ibid., 54.

[103] Ibid., 55.

[104] Ibid., 57.

[105] Ibid., 58.

[106] Ibid., 59–60.

[107] Pinczuk, Deane, and Kirkpatrick, *Sri Lanka (1976–2009)*, 196.

[108] Ethan Bueno de Mesquita, "The Quality of Terror," *American Journal of Political Science* 49, no. 3 (2005): 515.

[109] Kalyvas, *The Logic of Violence in Civil War*, 154.

[110] Ibid.

111 Luke N. Condra, Joseph H. Felter, Radha K. Iyengar, and Jacob N. Shapiro, "The Effect of Civilian Casualties in Afghanistan and Iraq" (working paper, National Bureau of Economic Research, Cambridge, MA, July 2010), http://www.nber.org/papers/w16152.

112 Ibid., 26.

113 Efraim Benmelech, Claude Berrebi, and Esteban Klor, "The Economic Cost of Harboring Terrorism," *Journal of Conflict Resolution* 54, no 2 (2010): 338, 340–345.

114 Ibid., 332.

115 Alberto Abadie and Javier Gardeazabal, "The Economic Costs of Conflict: A Case Study of the Basque Conflict," *The American Economic Review* 93, no. 1 (2003): 127.

116 Efraim Benmelech, Claude Berrebi, and Esteban Klor, "Economic Conditions and the Quality of Suicide Terrorism," *Journal of Politics* 74, no. 1 (2012): 121.

117 Christopher Paul, Colin P. Clarke, and Beth Grill, *Victory Has a Thousand Fathers: Sources of Success in Counterinsurgency* (Santa Monica, CA: RAND Corporation, 2010), 53.

118 Kalyvas, *The Logic of Violence in Civil War*, 150.

119 Ibid.

120 Ibid.

121 Ibid.

122 Ibid., 158–159.

123 Bronner, "Parsing Gains of Gaza War."

124 Paul, Clarke, and Grill, *Victory Has a Thousand Fathers*, 53.

125 Ibid., 23.

126 Paul, Clarke, and Grill, *Victory Has a Thousand Fathers*, xxvi.

127 Max Abrahms, "The Political Effectiveness of Terrorism Revisited," *Comparative Political Studies* 45, no. 3 (2012): 366–393.

128 Kalyvas, *The Logic of Violence in Civil War*, 152.

129 Ibid., 151.

130 Alexander B. Downes, "Draining the Sea by Filling the Graves: Investigating the Effectiveness of Indiscriminate Violence as a Counterinsurgency Strategy," *Civil Wars* 9, no. 4 (2007): 421.

131 Reed M. Wood, "Rebel Capability and Strategic Violence against Civilians," *Journal of Peace Research* 47, no. 5 (2010): 604.

132 Ibid., 605.

133 Kalyvas, *The Logic of Violence in Civil War*, 159.

134 Ibid., 165.

135 Ibid., 167–168.

136 Downes, "Draining the Sea by Filling the Graves," 422.

137 Ibid., 427, 436.

138 Ibid., 427.

139 Ibid., 439

140 Ibid., 422.

141 Ibid., 427, 439.

142 Nicholas Blanford, *Warriors of God: Inside Hezbollah's Thirty-Year Struggle against Israel* (Random House: New York, 2011), 42.

143 Luke N. Condra and Jacob N. Shapiro, "Who Takes the Blame? The Strategic Effects of Collateral Damage," *American Journal of Political Science* 56, no. 1 (2012): 167–187.

144 Ibid., 176.

145 Condra and Shapiro, "Who Takes the Blame?" 177–178.

146 Ibid., 180–181.

[147] David A. Jaeger, Esteban F. Klor, Sami H. Miaari, and M. Daniele Paserman, "The Struggle for Palestinian Hearts and Minds: Violence and Public Opinion in the Second Intifada," 355.

[148] Atkinson and Kress, "On Popular Response to Violence during Insurgencies," 224–226.

[149] Bloom, "Palestinian Suicide Bombing," 71.

[150] Blanford, *Warriors of God*, 82.

[151] Judith Harik, "Hizballah's Public and Social Services and Iran," in *Distant Relations: Iran and Lebanon in the Last 500 Years*, ed. H. E. Chehabi (London: I. B. Taurus & Co., 2006), 275.

[152] Ibid.

[153] Ibid., 275–276.

[154] Ibid., 278.

[155] Marc R. DeVore, "Exploring the Iran-Hezbollah Relationship: A Case Study of How State Sponsorship Affects Terrorist Group Decision-Making," *Perspectives on Terrorism* 6, no. 4–5 (2012): 94.

[156] Judith Harik, "Hizballah's Public and Social Services and Iran", 282.

[157] "Hezbollah Heartlands Recover with Iran's Help," *BBC News*, June 12, 2013, http://www.bbc.com/news/world-middle-east-22878198.

[158] Harik, "Hizballah's Public and Social Services and Iran," 273.

[159] Ibid.

[160] Ibid.

[161] Agan, *Narratives and Competing Messages*, 8–9.

[162] Jeffrey R. Halverson, H. Lloyd Goodall, and Steven R. Corman, *Master Narratives of Islamist Extremism* (New York: Palgrave McMillan, 2011), 1.

[163] Jerome Bruner, *The Culture of Education* (Cambridge, MA: Harvard University Press, 1996), 142.

[164] Ibid.; Wallace Martin, *Recent Theories of Narrative* (Ithaca, NY: Cornell University Press, 1996); Claude Levi-Strauss, *Structural Anthropology* (New York: Basic Books, 1967); Vladimir Propp, *Morphology of the Folktale* (Austin, TX: University of Texas Press, 1968); Victor Shklovsky, *Theory of Prose* (Elmwood Park, IL: Dalkey, 1990); Northrop Frye, *Anatomy of Criticism* (Princeton, NJ: Princeton University Press, 1957); W. J. T. Mitchell, *On Narrative* (Chicago: University of Chicago Press, 1981).

[165] Mark A. Finlayson and Steven R. Corman, "The Military Case for Narrative" (forthcoming).

[166] Erving Goffman, *Frame Analysis: An Essay on the Organization of Experience* (New York: Harper Colophon, 1974), 21.

[167] Robert D. Benford and David A. Snow, "Framing Processes and Social Movements: An Overview and Assessment," *Annual Review of Sociology* 26 (2000): 614.

[168] Ibid., 616–617.

[169] David A. Snow and Robert D. Benford, "Ideology, Frame Resonance, and Participant Mobilization," *International Social Movement Research* 1, no. 1 (1988): 198.

[170] Benford and Snow, "Framing Processes and Social Movements," 617.

[171] Ibid., 616–617.

[172] "The Hizballah Program: An Open Letter [to the Downtrodden in Lebanon and the World]," February 16, 1985, http://www.cfr.org/terrorist-organizations-and-networks/open-letter-hizballah-program/p30967.

[173] Halverson, Goodall, and Corman, *Master Narratives of Islamist Extremism*, 14.

[174] Daniel Leonard Bernardi, Pauline Hope Cheong, Christ Lundry, and Scott W. Ruston, *Narrative Landmines: Rumors Islamic Extremism, and the Struggle for Strategic Influence* (New Brunswick, NJ: Rutgers University Press, 2012), 36.

[175] Ibid., 34.

[176] Ibid., 35–36.

[177] "AQLIM Video Warns of Western Plots Against Africa, Urges Support for 'Mujahidin,'" Jihadist Websites (Al-Andalus Establishment for Media Production), March 9, 2010.

[178] Halverson, Goodall, and Corman, *Master Narratives of Islamist Extremism*, 149.

[179] "The Denudation of The Exoneration: Part 8," *Jihadica*, November 28, 2008, http://www.jihadica.com/the-denudation-of-the-exoneration-part-8/.

[180] Emmanuel Karagiannis, "Hezbollah as a Social Movement Organization: A Framing Approach," *Mediterranean Politics Journal* 14, no. 3 (2009): 376.

[181] Rudoren and Barnard, "Hezbollah Kills Two Israeli Soldiers Near Lebanon."

[182] Haggay Ram, *Myth and Mobilization in Revolutionary Iran: The Use of the Friday Congregational Sermon* (Washington, DC: American University Press, 1994), 71.

[183] Ibid.

[184] Ibid.

[185] Ibid., 45.

[186] Ibid.

[187] Bernardi et al., *Narrative Landmines*, 74.

[188] Ibid.

[189] Ibid., 88.

[190] Ibid., 84.

[191] Michael Billig, *Social Psychology and Intergroup Relations* (London: Academic Press, 1976), 343–352.

[192] Donald Horowitz, *Ethnic Groups in Conflict* (Berkeley: University of California Press, 1985), 143.

[193] Roger D. Petersen, *Understanding Ethnic Violence: Fear, Hatred, and Resentment in Twentieth-Century Eastern Europe* (Cambridge, UK: Cambridge University Press, 2002).

[194] Ibid., 128–129.

[195] Ibid., 121–123.

[196] Ibid., 124.

[197] Ibid., 125.

[198] Ibid., 127.

[199] Ibid., 129.

[200] Ted Gurr, "A Causal Model of Civil Strife: A Comparative Analysis Using New Indices," *The American Political Science Review* 62, no. 4 (1968): 1104

[201] Jeremy Ginges, Scott Atran, Douglas Medin, and Khalil Shikaki, "Sacred Bounds on Rational Resolution of Violent Political Conflict," *Proceedings of the National Academy of Sciences* 104, no. 18 (2007): 7357–7360.

[202] Ibid., 7358.

[203] Ibid., 7358–7359.

[204] Ibid., 7359–7360.

[205] Ibid.

[206] Daniel Kahneman and Amos Tversky, "Prospect Theory: An Analysis of Decision under Risk," *Econometrica* 47, no. 2 (1979): 263–292.

[207] Rose McDermott, *Risk-Taking in International Politics: Prospect Theory in American Foreign Policy* (Ann Arbor: University of Michigan Press, 1998), 21.

[208] Ibid., 18.

[209] Ibid., 47–51.

[210] Eli Berman, Jacob N. Shapiro, and Joseph H. Felter, "Can Hearts and Minds Be Bought? The Economics of Counterinsurgency in Iraq," *Journal of Political Economy* 119, no. 4 (2011): 781, 814; Bueno de Mesquita and Dickson, "The Propaganda of the Deed," 372–374; Bueno de Mesquita, "The Quality of Terror."

[211] Office of General Counsel, Department of Defense, "Department of Defense Law of War Manual," June 2015, http://archive.defense.gov/pubs/law-of-war-manual-june-2015.pdf.

[212] Kalyvas, *The Logic of Violence in Civil War*, 167.

[213] DeNardo, *Power in Numbers*, 202.

[214] Efraim Benmelech, Claude Berrebi, and Esteban Klor, "Counter-Suicide Terrorism: Evidence from House Demolitions" (working paper, National Bureau of Economic Research, Cambridge, MA, October 2010), 2.

[215] Condra and Shapiro, "Who Takes the Blame?" 178; more specifically, see table SE Table 2G in the online appendix to the article.

[216] Condra, Felter, Iyengar, and Shapiro, "The Effect of Civilian Casualties in Afghanistan and Iraq," 7–9.

BIBLIOGRAPHY

Abadie, Alberto, and Javier Gardeazabal. "The Economic Costs of Conflict: A Case Study of the Basque Conflict." *The American Economic Review* 93, no. 1 (2003): 113–132.

Abrahms, Max, and Philip B. K. Potter. "Explaining Terrorism: Leadership Deficits and Militant Group Tactics." *International Organization* 69, no. 2 (2015): 311–342.

Abrahms, Max. "The Political Effectiveness of Terrorism Revisited." *Comparative Political Studies* 45, no. 3 (2012): 366–393.

Agan, Summer. *Narratives and Competing Messages.* Fort Bragg, NC: United States Army Special Operations Command, forthcoming.

Anderson, Benedict. *Imagined Communities.* London: Verso, 2006.

"AQLIM Video Warns of Western Plots Against Africa, Urges Support for 'Mujahidin,'" Jihadist Websites (Al-Andalus Establishment for Media Production), March 9, 2010.

Atkinson, Michael P., and Moshe Kress. "On Popular Response to Violence during Insurgencies." *Operations Research Letters* 40, no 4 (2012): 223–229.

"Back to Bashing: Messrs. Netanyahu and Nasrallah Flirt with War Once More," *The Economist,* January 31, 2015, http://www.economist.com/news/middle-east-and-africa/21641308-messrs-netanyahu-and-nasrallah-flirt-war-once-more-back-bashing.

Bandarage, Asoka. *The Separatist Conflict in Sri Lanka: Terrorism, Ethnicity, Political Economy.* New York and Bloomington, IN: iUniverse, 2009.

Benford, Robert D., and David A. Snow. "Framing Processes and Social Movements: An Overview and Assessment." *Annual Review of Sociology* 26 (2000): 611–639.

Benmelech, Efraim, Claude Berrebi, and Esteban Klor. "Counter-Suicide Terrorism: Evidence from House Demolitions." NBER Working Paper No. 16493, National Bureau of Economic Research, Cambridge, MA, October 2010.

Benmelech, Efraim, Claude Berrebi, and Esteban Klor. "Economic Conditions and the Quality of Suicide Terrorism." *Journal of Politics* 74, no 1 (2012): 113–128.

Benmelech, Efraim, Claude Berrebi, and Esteban Klor. "The Economic Cost of Harboring Terrorism." *Journal of Conflict Resolution* 54, no 2 (2010): 331–353.

Berman, Eli, Jacob N. Shapiro, and Joseph H. Felter. "Can Hearts and Minds Be Bought? The Economics of Counterinsurgency in Iraq." *Journal of Political Economy* 119, no. 4 (2011): 766–819.

Bernardi, Daniel Leonard, Pauline Hope Cheong, Christ Lundry, and Scott W. Ruston. *Narrative Landmines: Rumors Islamic Extremism, and the Struggle for Strategic Influence*. New Brunswick, NJ: Rutgers University Press, 2012.

Billig, Michael. *Social Psychology and Intergroup Relations*. London: Academic Press, 1976.

Blainey, Geoffrey. *The Causes of War*, 3rd ed. New York: Free Press, 1988.

Blanford, Nicholas. *Warriors of God: Inside Hezbollah's Thirty-Year Struggle against Israel*. Random House: New York, 2011.

Bloom, Mia. "Palestinian Suicide Bombing: Public Support, Market Share and Outbidding." *Political Science Quarterly* 119, no. 1 (Spring 2004): 61–88.

Bronner, Ethan. "Parsing Gains of Gaza War," *New York Times*, January 19, 2009, http://www.nytimes.com/2009/01/19/world/middleeast/19assess.html?hp=&_r=0.

Bruner, Jerome. *The Culture of Education*. Cambridge, MA: Harvard University Press, 1996.

Buikema, Ron, and Matt Burger. "Sendero Luminoso (Shining Path)." In *Casebook on Insurgency and Revolutionary Warfare Volume II: 1962–2009*. Edited by Chuck Crossett. Ft. Bragg, NC: United States Army Special Operations Command, 2012.

Bueno de Mesquita, Ethan, and Eric S. Dickson. "The Propaganda of the Deed: Terrorism, Counterterrorism, and Mobilization." *American Journal of Political Science* 51, no. 2 (2007): 364–381.

Bueno de Mesquita, Ethan. "The Quality of Terror." *American Journal of Political Science* 49, no. 3 (2005): 515–530.

Byman, Daniel. "The Logic of Ethnic Terrorism." *Studies in Conflict & Terrorism* 21, no. 2 (1998): 149–169.

Caplan, Bryan. "Terrorism: The Relevance of the Rational Choice Model." *Public Choice* 128, no. 1–2 (2006): 91–107.

Chandraprema, C. A. *Sri Lanka: The Years of Terror. The JVP Insurrection 1987–89*. Colombo, Sri Lanka: Lake House Bookshop, 1991.

Cliffe, Lionel, Joshua Mpofu, and Barry Munslow. "Nationalist Politics in Zimbabwe: The 1980 Elections and Beyond." Review of African Political Economy 7, no. 18 (1980): 55.

Collier, Paul, and Anke Hoeffler. "Greed and Grievance in Civil War." *Oxford Economic Papers* 56, no. 4 (2004): 563–595.

Condra, Luke N., and Jacob N. Shapiro. "Who Takes the Blame? The Strategic Effects of Collateral Damage." *American Journal of Political Science* 56, no. 1 (2012): 167–187.

Condra, Luke N., Joseph H. Felter, Radha K. Iyengar, and Jacob N. Shapiro. "The Effect of Civilian Casualties in Afghanistan and Iraq." NBER Working Paper No. 16152, National Bureau of Economic Research, Cambridge, MA, July 2010, http://www.nber.org/papers/w16152.

Darby, John. "Legitimate Targets: A Control on Violence?" In *New Perspectives on the Northern Ireland Conflict.* Edited by Adrian Guelke. Aldershot, England: Avebury, 1994.

DeNardo, James. *Power in Numbers: The Political Strategy of Protest and Rebellion.* Princeton, NJ: Princeton University Press, 1985.

DeVore, Marc R. "Exploring the Iran-Hezbollah Relationship: A Case Study of How State Sponsorship Affects Terrorist Group Decision-Making." *Perspectives on Terrorism* 6, no. 4–5 (2012): 85–107.

Diplomatic Correspondent. "Rajiv Assassination 'Deeply Regretted': LTTE," *The Hindu,* June 28, 2006, http://www.thehindu.com/todays-paper/rajiv-assassination-quotdeeply-regretted-ltte/article3125569.ece.

Downes, Alexander B. "Draining the Sea by Filling the Graves: Investigating the Effectiveness of Indiscriminate Violence as a Counterinsurgency Strategy." *Civil Wars* 9, no. 4 (2007): 420–444.

Drake, C. J. M. "The Role of Ideology in Terrorists' Target Selection." *Terrorism and Political Violence* 10, no. 2 (1998): 53–85.

Finlayson, Mark A., and Steven R. Corman. "The Military Case for Narrative." Forthcoming.

Frye, Northrop. *Anatomy of Criticism.* Princeton, NJ: Princeton University Press, 1957.

Ginges, Jeremy, Scott Atran, Douglas Medin, and Khalil Shikaki. "Sacred Bounds on Rational Resolution of Violent Political Conflict." *Proceedings of the National Academy of Sciences* 104, no. 18 (2007): 7357–7360.

Goffman, Erving. *Frame Analysis: An Essay on the Organization of Experience.* New York: Harper Colophon, 1974.

Green, Donald P., and Ian Shapiro. *Pathologies of Rational Choice Theory: A Critique of Applications in Political Science.* New Haven, CT: Yale University Press, 1994.

Gurr, Ted. *"A Causal Model of Civil Strife: A Comparative Analysis Using New Indices."* American Political Science Review 62, no. 4 (1968): 1104-1124.

Halverson, Jeffrey R., H. Lloyd Goodall, and Steven R. Corman. *Master Narratives of Islamist Extremism.* New York: Palgrave McMillan, 2011.

Harik, Judith. "Hizballah's Public and Social Services and Iran." In *Distant Relations: Iran and Lebanon in the Last 500 Years.* Edited H. E. Chehabi. London: I. B. Taurus & Co., 2006.

"Hezbollah Heartlands Recover with Iran's Help," *BBC News,* June 12, 2013, http://www.bbc.com/news/world-middle-east-22878198.

Horowitz, Donald. *Ethnic Groups in Conflict.* Berkeley: University of California Press, 1985.

Jaeger, David A., Esteban F. Klor, Sami H. Miaari, and M. Daniele Paserman. "Can Militants Use Violence to Win Public Support? Evidence from the Second Intifada." *Journal of Conflict Resolution* 59, no. 3 (2015): 528–549.

———. "The Struggle for Palestinian Hearts and Minds: Violence and Public Opinion in the Second Intifada." *Journal of Public Economics* 96, nos. 3-4 (2012): 354–368.

Jehl, Douglas, and Thom Shanker. "The Struggle for Iraq: Terrorist Liaisons; Al Qaeda Tells Ally in Iraq to Strive for Global Goals," *New York Times,* October 7, 2005, http://www.nytimes.com/2005/10/07/world/the-struggle-for-iraq-terrorist-liaisons-al-qaeda-tells-ally-in-iraq-to-strive-for-global-goals.html?_r=0.

Jones, Bryan D. "Bounded Rationality." *Annual Review of Political Science* 2 (1999): 297–321.

Kahneman, Daniel, and Amos Tversky. "Prospect Theory: An Analysis of Decision under Risk." *Econometrica* 47, no. 2 (1979): 263–292.

Kalyvas, Stathis N. *The Logic of Violence in Civil War.* Cambridge, UK: Cambridge University Press, 2006.

Karagiannis, Emmanuel. "Hezbollah as a Social Movement Organization: A Framing Approach." *Mediterranean Politics Journal* 14, no. 3 (2009): 365–383.

Kaufmann, Chaim. "Possible and Impossible Solutions to Ethnic Civil Wars." *International Security* 20, no. 4 (1996): 136–175.

Kuznar, Lawrence A. "Rationality Wars and the War on Terror: Explaining Terrorism and Social Unrest." *American Anthropologist* 109, no. 2 (2007): 318–329.

Kydd, Andrew H., and Barbara F. Walter. "The Strategies of Terrorism." *International Security* 31, no. 1 (2006): 49–80.

Leites, Nathan, and Charles Wolf Jr.. *Rebellion and Authority: An Analytic Essay on Insurgent Conflicts.* Chicago: Markham Publishing Company, 1970.

Levi-Strauss, Claude. *Structural Anthropology.* New York: Basic Books, 1967.

Libicki, Martin C., Peter Chalk, and Melanie Sisson. *Exploring Terrorist Targeting Preferences.* Santa Monica, CA: RAND Corporation, 2007.

Martin, Wallace. *Recent Theories of Narrative.* Ithaca, NY: Cornell University Press, 1996.

Mason, T. David. "Insurgency, Counterinsurgency, and the Rational Peasant." *Public Choice* 86, no. 1–2 (1996): 63–83.

McCary, John A. "The Anbar Awakening: An Alliance of Incentives." *Washington Quarterly* 32, no. 1 (2009): 43–59.

McCormick, Gordon H. "Terrorist Decision Making." *Annual Review of Political Science* 6, (2003): 473–507.

McCormick, Gordon H., and Frank Giordano. "Things Come Together: Symbolic Violence and Guerrilla Mobilisation." *Third World Quarterly* 28, no. 2 (2007): 295–320.

McDermott, Rose. *Risk-Taking in International Politics: Prospect Theory in American Foreign Policy.* Ann Arbor: University of Michigan Press, 1998.

Mitchell, W. J. T. *On Narrative.* Chicago: University of Chicago Press, 1981.

Moore, Mick. "Thoroughly Modern Revolutionaries: The JVP in Sri Lanka." *Modern Asian Studies* 27, no. 3 (July 1993): 593–642.

Nemeth, Stephen Charles. "A Rationalist Explanation of Terrorist Targeting." PhD diss., University of Iowa, 2010, http://ir.uiowa.edu/etd/718/.

Office of General Counsel, Department of Defense, "Department of Defense Law of War Manual," June 2015, http://archive.defense.gov/pubs/law-of-war-manual-june-2015.pdf.

Olson, Mancur. *The Logic of Collective Action: Public Goods and the Theory of Groups.* Cambridge, MA: Harvard University Press, 2009.

Overgaard, Per Baltzer. "The Scale of Terrorist Attacks as a Signal of Resources." *Journal of Conflict Resolution* 38, no. 3 (1994): 452–478.

Paul, Christopher, Colin P. Clarke, and Beth Grill. *Victory Has a Thousand Fathers: Sources of Success in Counterinsurgency.* Santa Monica, CA: RAND Corporation, 2010.

Pedahzur, Ami. *Suicide Terrorism.* Cambridge, UK: Polity Press, 2005.

Petersen, Roger D. *Understanding Ethnic Violence: Fear, Hatred, and Resentment in Twentieth-Century Eastern Europe.* Cambridge, UK: Cambridge University Press, 2002.

Pinczuk, Guillermo, Mike Deane, and Jesse Kirkpatrick. *Case Studies in Insurgency and Revolutionary Warfare—Sri Lanka (1976–2009).* Edited by Guillermo Pinczuk. Fort Bragg, NC: United States Army Special Operations Command, 2014.

Popkin, Samuel L. *The Rational Peasant: The Political Economy of Rural Society in Vietnam.* Berkeley: University of California Press, 1979.

Propp, Vladimir. *Morphology of the Folktale.* Austin, TX: University of Texas Press, 1968.

Ram, Haggay. *Myth and Mobilization in Revolutionary Iran: The Use of the Friday Congregational Sermon.* Washington, DC: American University Press, 1994.

Richardson, Peter J. and Robert Boyd. Not by Genes Alone: How Culture Transformed Human Evolution. Chicago: University of Chicago Press, 2005.

Rudoren, Jodi, and Anne Barnard. "Hezbollah Kills Two Israeli Soldiers Near Lebanon," *New York Times,* January 28, 2015, http://www.nytimes.com/2015/01/29/world/middleeast/israel-lebanon-hezbollah-missile-attack.html?_r=0.

Shklovsky, Victor. *Theory of Prose.* Elmwood Park, IL: Dalkey, 1990.

Snow, David A., and Robert D. Benford. "Ideology, Frame Resonance, and Participant Mobilization." *International Social Movement Research* 1, no. 1 (1988): 197–217.

Staniland, Paul. "States, Insurgents, and Wartime Political Orders." *Perspectives on Politics* 10, no. 2 (2012): 243–264.

Stephan, Maria J., and Erica Chenoweth. "Why Civil Resistance Works: The Strategic Logic of Nonviolent Conflict." *International Security* 33, no. 1 (2008): 7–44.

"The Denudation of The Exoneration: Part 8," *Jihadica,* November 28, 2008, http://www.jihadica.com/the-denudation-of-the-exoneration-part-8/.

"The Hizballah Program: An Open Letter [to the Downtrodden in Lebanon and the World]," February 16, 1985, http://www.cfr.org/terrorist-organizations-and-networks/open-letter-hizballah-program/p30967.

Thomas, Jakana. "Rewarding Bad Behavior: How Governments Respond to Terrorism in Civil War." *American Journal of Political Science* 58, no. 4 (2014): 804–818.

Thornton, Thomas. "Terror as a Weapon of Political Agitation." In *Internal War.* Edited by Harry Eckstein. Westport, CT: Greenwood Press, 1980.

Walt, Stephen M. "Rigor or Rigor Mortis? Rational Choice and Security Studies." *International Security* 23, no. 4 (1999): 5-48.

"War By Any other Name," *The Economist,* July 5, 2014, http://www.economist.com/news/europe/21606290-russia-has-effect-already-invaded-eastern-ukraine-question-how-west-will.

Weinstein, Jeremy M. *Inside Rebellion: The Politics of Insurgent Violence.* Cambridge: Cambridge University Press, 2007.

Wendt, Eric P. "Strategic Counterinsurgency Modeling." *Special Warfare* 18, no. 2 (2005): 2–13.

Wood, Reed M. "Rebel Capability and Strategic Violence against Civilians." Journal of Peace Research 47, no. 5 (2010): 604.

York, Anthony, "Israel Rally Critical of Bush." Salon.com, April 16, 2002.

INDEX

www.ingramcontent.com/pod-product-compliance
Lightning Source LLC
Chambersburg PA
CBHW052117020426
42335CB00021B/2801